MY FAVOURITE DOG STORIES

Also in this series

MY FAVOURITE ANIMAL STORIES
edited by Gerald Durrell

MY FAVOURITE STORIES OF EXPLORATION
edited by David Attenborough

MY FAVOURITE STORIES OF WILD LIFE
edited by Peter Scott

MY FAVOURITE COUNTRY STORIES
edited by Henry Williamson

MY FAVOURITE STORIES OF TRAVEL
edited by Eric Newby

MY FAVOURITE HORSE STORIES
edited by Dorian Williams

MY FAVOURITE SEA STORIES
edited by Alan Villiers

My Favourite
DOG STORIES

edited by

DOUGLAS BADER

With drawings by

VICTOR CAESARI

LUTTERWORTH PRESS
GUILDFORD AND LONDON

First published in this collected form 1969
Reprinted 1972

ISBN 0 7188 1360 x

Printed in Great Britain by
Compton Printing Ltd., London and Aylesbury

Acknowledgments

The editor is indebted to the following for permission to include material which is their copyright:

Macmillan & Co. Ltd. and Mrs. George Bambridge, for "Garm—A Hostage", from *Actions and Reactions* by Rudyard Kipling

Methuen & Co., Ltd., for "Tatters and Scraps—Kennel-Terriers", from *Maria and Some Other Dogs* by E. Œ. Somerville and Martin Ross

Longmans, Green & Co., Ltd., for "The Pick of the Puppies", from *Jock of the Bushveld* by Sir Percy Fitzpatrick

Messrs. William Heinemann Ltd., for "Rex", from *Phoenix* by D. H. Lawrence

Hamish Hamilton, Ltd., for "Josephine Has Her Day", from *Thurber's Dogs* by James Thurber

Rupert Hart-Davis Ltd., for "Dodo", from *My Family and Other Animals* by Gerald Durrell

The Hogarth Press Ltd., and Mr. Leonard Woolf, for "Flush", from *Flush* by Virginia Woolf;
also for "For the Love of a Man", from *The Call of the Wild* by Jack London.

George G. Harrap & Co., Ltd., for "The Training of Rex", from *My Dog Rex* by Arthur Holman

Contents

Introduction 9

1. GARM—A HOSTAGE 11
from *Actions and Reactions* by Rudyard Kipling

2. TATTERS AND SCRAPS 31
from *Maria and Some Other Dogs,* by E. Œ. Somerville and Martin Ross

3. THE PICK OF THE PUPPIES 38
from *Jock of the Bushveld,* by Sir Percy Fitzpatrick

4. REX 52
from *The Posthumous Papers of D. H. Lawrence*

5. JOSEPHINE HAS HER DAY 62
from *Thurber's Dogs,* by James Thurber

6. DODO 80
from *My Family and Other Animals* by Gerald Durrell

7. FLUSH 88
from *Flush* by Virginia Woolf

8. FOR THE LOVE OF A MAN 113
from *The Call of the Wild* by Jack London

9. TRAINING A POLICE DOG 128
from *My Dog Rex* by Arthur Holman

Introduction

Being an Englishman, I was brought up in the tradition of a love of animals, especially horses and dogs; however I found it difficult to select stories for this book for when one thinks of the many stories about dogs which we all know and love—who am I to pick and choose! Yet that is exactly what I have done.

Personally, I like reading about dogs who are big and strong, courageous and loyal, like Buck in "For the Love of a Man". It is a chapter from *The Call of the Wild* by that magnificent writer Jack London, and is but an episode in Buck's life. What a dog! half St. Bernard and half Scottish sheepdog. What a mixture!

Many readers will enjoy "Josephine Has Her Day". Not quite a dog story in the literal sense, it is more about a dog's influence on a rather mild-mannered man. James Thurber tells it well.

Young readers will enjoy "Rex" by D. H. Lawrence.

"The Pick of the Puppies" is almost nostalgically true to life, and enchantingly told, and "Tatters and Scraps—Kennel-Terriers" is charming and typical of the subject.

The spaniel Flush changed her way of life from country to town dog, to become the companion of the famous Elizabeth Barrett Browning. This brought Flush fame in book and in film. Virginia Woolf writes the story in unusual and fascinating style by telling it from Flush's point of view.

I began with Buck: let me finish with another of the same type. Garm—that magnificent bull terrier who was intellectually human, and truly loyal to one man. Like most of Rudyard Kipling's works, the background is British India at the time of the Empire.

Some readers of this book may wonder at the omission of their own favourite dog story. The reason is clear—no book would be big enough to hold them all. My own favourite reminiscences are about Shaun, our golden retriever, whom my wife was given when he was six weeks old, in 1941. He died aged eleven, in 1952. Golfer, swimmer, aviator, he ruled our lives while he lived, and broke our hearts when he died.

Let me finish by quoting the final sentence in the book *My Dog Rex,* a chapter of which forms the last story in this book:

"Ours was no ordinary 'master and pet' relationship. Police Dog Rex III was an essential part of my life and being, and still dominates my thoughts and memories. When he died something vital went out of me, for in Rex I really did 'give my heart to a dog to tear'."

No finer tribute was ever paid by a man to his dog.

I *Garm—a Hostage*

RUDYARD KIPLING

One night, a very long time ago, I drove to an Indian military cantonment called Mian Mir to see amateur theatricals. At the back of the Infantry barracks a soldier, his cap over one eye, rushed in front of the horses and shouted that he was a dangerous highway robber. As a matter of fact he was a friend of mine, so I told him to go home before any one caught him; but he fell under the pole, and I heard voices of a military guard in search of someone.

The driver and I coaxed him into the carriage, drove home swiftly, undressed him and put him to bed, where he waked next morning with a sore headache, very much ashamed. When his uniform was cleaned and dried, and he had been shaved and washed and made neat, I drove him back to barracks with his arm in a fine white sling, and reported that I had accidentally run over him. I did not tell this story to my friend's sergeant, who was a hostile and unbelieving person, but to his lieutenant, who did not know us quite so well.

Three days later my friend came to call, and at his heels slobbered and fawned one of the finest bull-terriers—of the old-fashioned breed, two parts bull and one terrier—that I had ever set eyes on. He was pure white, with a fawn-coloured saddle just behind his neck, and a fawn diamond at the root of his thin whippy tail. I had admired him distantly for more than a year; and Vixen, my own fox-terrier, knew him too, but did not approve.

" 'E's for you", said my friend; but he did not look as though he liked parting with him.

"Nonsense! That dog's worth more than most men, Stanley", I said.

" 'E's that an' more. 'Tention!"

The dog rose on his hind legs, and stood upright for a full minute.

"Eyes right!"

He sat on his haunches and turned his head sharp to the right. At a sign he rose and barked thrice. Then he shook hands with his right paw and bounded lightly to my shoulder. Here he made himself into a necktie, limp and lifeless, hanging down on either side of my neck. I was told to pick him up and throw him in the air. He fell with a howl, and held up one leg.

"Part o' the trick", said his owner. "You're going to die now. Dig yourself your little grave an' shut your little eye".

Still limping, the dog hobbled to the garden-edge, dug a hole and lay down in it. When told that he was cured he jumped out, wagging his tail, and whining for applause. He was put through half a dozen other tricks, such as showing how he would hold a man safe (I was that man, and he sat down before me, his teeth bared, ready to spring), and how he would cease eating at the word of command. I had no more than finished praising him when my friend made a gesture that stopped the dog as though he had been shot, took a piece of blue-ruled canteen-paper from his helmet, handed it to me and ran away, while the dog looked after him and howled. I read:—

> SIR—I give you the dog because of what you got me out of. He is the best I know, for I made him myself, and he is as good as a man. Please do not give him too much to eat, and please do not give him back to me, for I'm not going to take him, if you will keep him. So please do not try to give him back any more. I have kept his name back, so you can call him anything and he will answer, but please do not give him back. He can kill a man as easy as anything, but please do not give him too much meat. He knows more than a man.

Vixen sympathetically joined her shrill little yap to the bull-terrier's despairing cry, and I was annoyed, for I knew that a man who cares for dogs is one thing, but a man who loves one dog is quite another. Dogs are, at the best, no more than verminous vagrants, self-scratchers, foul feeders, and unclean by the law of Moses and Mohammed; but a dog with whom one lives alone for at least six months in the year; a free thing, tied to you so strictly by love that without you he will not stir or exercise; a patient, temperate, humorous, wise soul, who knows your moods before you know them yourself, is not a dog under any ruling.

I had Vixen, who was all my dog to me; and I felt what my friend must have felt, at tearing out his heart in this style and leaving it in my garden. However, the dog understood clearly enough that I was his master, and did not follow the soldier. As soon as he drew breath

I made much of him, and Vixen, yelling with jealousy, flew at him. Had she been of his own sex, he might have cheered himself with a fight, but he only looked worriedly when she nipped his deep iron sides, laid his heavy head on my knee, and howled anew. I meant to dine at the Club that night, but as darkness drew in, and the dog snuffed through the empty house like a child trying to recover from a fit of sobbing, I felt that I could not leave him to suffer his first evening alone. So we fed at home, Vixen on one side and the stranger-dog on the other; she watching his every mouthful, and saying explicitly what she thought of his table manners, which were much better than hers.

It was Vixen's custom, till the weather grew hot, to sleep in my bed, her head on the pillow like a Christian; and when morning came I would always find that the little thing had braced her feet against the wall and pushed me to the very edge of the cot. This night she hurried to bed purposefully, every hair up and one eye on the stranger, who had dropped on a mat in a helpless, hopeless sort of way, his four feet spread out, sighing heavily. She settled her head on the pillow several times, to show her little airs and graces, and struck up her usual whiney sing-song before slumber. The stranger-dog softly edged towards me. I put out my hand and he licked it. Instantly my wrist was between Vixen's teeth, and her warning *aaarh!* said as plainly as speech that if I took any further notice of the stranger she would bite.

I caught her behind her fat neck with my left hand, shook her severely, and said:—

"Vixen, if you do that again you'll be put into the veranda. Now, remember!"

She understood perfectly, but the minute I released her she mouthed my right wrist once more, and waited with her ears back and all her body flattened, ready to bite. The big dog's tail thumped the floor in a humble and peace-making way.

I grabbed Vixen a second time, lifted her out of bed like a rabbit (she hated that and yelled), and, as I had promised, set her out in the veranda with the bats and the moonlight. At this she howled. Then she used coarse language—not to me, but to the bull-terrier—till she coughed with exhaustion. Then she ran round the house trying every door. Then she went off to the stables and barked as though someone were stealing the horses, which was an old trick of hers. Last she returned, and her snuffing yelp said, "I'll be good! Let me in and I'll be good!"

She was admitted and flew to her pillow. When she was quieted I whispered to the other dog, "You can lie on the foot of the bed." The bull jumped up at once, and though I felt Vixen quiver with rage, she knew better than to protest. So we slept till the morning, and they had early breakfast with me, bite for bite, till the horse came round and we went for a ride. I don't think the bull had ever followed a horse before. He was wild with excitement, and Vixen, as usual, squealed and scuttered and scooted, and took charge of the procession.

There was one corner of a village near by, which we generally passed with caution, because all the yellow pariah-dogs of the place gathered about it. They were half-wild, starving beasts, and though utter cowards, yet where nine or ten of them get together they will mob and kill and eat an English dog. I kept a whip with a long lash for them. That morning they attacked Vixen, who, perhaps of design, had moved from beyond my horse's shadow.

The bull was ploughing along in the dust, fifty yards behind, rolling in his run, and smiling as bull-terriers will. I heard Vixen squeal. Half a dozen of the curs closed in on her; a white streak came up behind me;

a cloud of dust rose near Vixen, and, when it cleared, I saw one tall pariah with his back broken, and the bull wrenching another to earth. Vixen retreated to the protection of my whip, and the bull paddled back smiling more than ever, covered with the blood of his enemies. That decided me to call him "Garm of the Bloody Breast," who was a great person in his time, or "Garm" for short; so, leaning forward, I told him what his temporary name would be. He looked up while I repeated it, and then raced away. I shouted "Garm!" He stopped, raced back, and came up to ask my will.

Then I saw that my soldier-friend was right, and that that dog knew, and was worth, more than a man. At the end of the ride I gave an order which Vixen knew and hated: "Go away and get washed!" I said. Garm understood some part of it, and Vixen interpreted the rest, and the two trotted off together soberly. When I went to the back veranda Vixen had been washed snowy-white, and was very proud of herself, but the dog-boy would not touch Garm on any account unless I stood by. So I waited while he was being scrubbed, and Garm, with the soap creaming on the top of his broad head, looked at me to make sure that this was what I expected him to endure. He knew perfectly that the dog-boy was only obeying orders.

"Another time," I said to the dog-boy, "you will wash the great dog with Vixen when I send them home."

"Does *he* know?" said the dog-boy, who understood the ways of dogs.

"Garm," I said, "another time you will be washed with Vixen."

I knew that Garm understood. Indeed, next washing-day, when Vixen as usual fled under my bed, Garm stared at the doubtful dog-boy in the veranda, stalked to the place where he had been washed last time, and stood rigid in his tub.

But the long days in my office tried him sorely. We three would drive off in the morning at half past eight and come home at six or later. Vixen, knowing the routine of it, went to sleep under my table; but the confinement ate into Garm's soul. He generally sat on the veranda looking out on the Mall; and well I knew what he expected.

Sometimes a company of soldiers would move along on their way to the Fort, and Garm rolled forth to inspect them; or an officer in uniform entered into the office, and it was pitiful to see poor Garm's welcome to the cloth—not the man. He would leap at him, and sniff and bark joyously, then run to the door and back again. One afternoon I heard him bay with a full throat —a thing he had never done before—and he disappeared. When I drove into my garden at the end of the day a soldier in white uniform scrambled over the wall at the far end, and the Garm that met me was a joyous dog. This happened twice or thrice a week for a month.

I pretended not to notice, but Garm knew and Vixen knew. He would glide homewards from the office about four o'clock, as though he were only going to look at the scenery, and this he did so quietly that but for Vixen I should not have noticed him. The jealous little dog under the table would give a sniff and a snort, just loud enough to call my attention to the flight. Garm might go out forty times in the day and Vixen would never stir, but when he slunk off to see his true master in my garden she told me in her own tongue. That was the one sign she made to show that Garm did not altogether belong to the family. They were the best of friends at all times, *but,* Vixen explained that I was never to forget Garm did not love me as she loved me.

I never expected it. The dog was not my dog—could

never be my dog—and I knew he was as miserable as his master who tramped eight miles a day to see him. So it seemed to me that the sooner the two were reunited the better for all. One afternoon I sent Vixen home alone in the dog-cart (Garm had gone before), and rode over to cantonments to find another friend of mine, who was an Irish soldier and a great friend of the dog's master.

I explained the whole case, and wound up with:

"And now Stanley's in my garden crying over his dog. Why doesn't he take him back? They're both unhappy."

"Unhappy! There's no sense in the little man any more. But 'tis his fit."

"What *is* his fit? He travels fifty miles a week to see the brute, and he pretends not to notice me when he sees me on the road; and I'm as unhappy as he is. Make him take the dog back."

"It's his penance he's set himself. I told him by way av a joke, afther you'd run over him so convenient that night, whin he was dhrunk—I said if he was a Catholic he'd do penance. Off he went wid that fit in his little head *an'* a dose av fever, an' nothin' would suit but givin' you the dog as a hostage."

"Hostage for what? I don't want hostages from Stanley."

"For his good behaviour. He's keepin' straight now, the way it's no pleasure to associate wid him."

"Has he taken the pledge?"

"If 'twas only *that* I need not care. Ye can take the pledge for three months on an' off. He sez he'll never see the dog again, an' *so*, mark you, he'll keep straight for evermore. Ye know his fits? Well, this is wan av them. How's the dog takin' it?"

"Like a man. He's the best dog in India. Can't you make Stanley take him back?"

"I can do no more than I have done. But ye know his fits. He's just doin' his penance. What will he do when he goes to the Hills? The docthor's put him on the list."

It is the custom in India to send a certain number of invalids from each regiment up to stations in the Himalayas for the hot weather; and though the men ought to enjoy the cool and the comfort, they miss the society of the barracks down below, and do their best to come back or to avoid going. I felt that this move would bring matters to a head, so I left Terence hopefully, though he called after me:—

"He won't take the dog, sorr. You can lay your month's pay on that. Ye know his fits."

I never pretended to understand Private Ortheris; and so I did the next best thing—I left him alone.

That summer the invalids of the regiment to which my friend belonged were ordered off to the Hills early, because the doctors thought marching in the cool of the day would do them good. Their route lay south to a place called Umballa, a hundred and twenty miles or more. Then they would turn east and march up into the hills to Kasauli or Dugshai or Subathoo. I dined with the officers the night before they left—they were marching at five in the morning. It was midnight when I drove into my garden and surprised a white figure flying over the wall.

"That man," said my butler, "has been here since nine, making talk to that dog. He is quite mad. I did not tell him to go away because he has been here many times before, and because the dog-boy told me that if I told him to go away, that great dog would immediately slay me. He did not wish to speak to the Protector of the Poor, and he did not ask for anything to eat or drink."

"Kadir Buksh," said I, "that was well done, for the dog would surely have killed thee. But I do not think the white soldier will come any more."

Garm slept ill that night and whimpered in his dreams. Once he sprang up with a clear, ringing bark, and I heard him wag his tail till it waked him and the bark died out in a howl. He had dreamed he was with his master again, and I nearly cried. It was all Stanley's silly fault.

The first halt which the detachment of invalids made was some miles from their barracks, on the Amritzar road, and ten miles distant from my house. By a mere chance one of the officers drove back for another good dinner at the Club (cooking on the line of march is always bad), and there I met him. He was a particular friend of mine, and I knew that he knew how to love a dog properly. His pet was a big fat retriever who was going up to the Hills for his health, and though it was still April, the round brown brute puffed and panted in the Club veranda as though he would burst.

"It's amazing," said the officer, "what excuses these invalids of mine make to get back to barracks. There's a man in my company now asked me for leave to go back to cantonments to pay a debt he'd forgotten. I was so taken by the idea I let him go, and he jingled off in an *ekka* as pleased as Punch. Ten miles to pay a debt! Wonder what it was really?"

"If you'll drive me home I think I can show you," I said.

So we went over to my house in his dog-cart with the retriever; and on the way I told him the story of Garm.

"I was wondering where that brute had gone to. He's the best dog in the regiment," said my friend. "I offered the little fellow twenty rupees for him a month ago. But he's a hostage, you say, for Stanley's good conduct. Stanley's one of the best men I have—when he chooses."

"That's the reason why," I said. "A second-rate man

wouldn't have taken things to heart as he has done."

We drove in quietly at the far end of the garden, and crept round the house. There was a place close to the wall all grown about with tamarisk trees, where I knew Garm kept his bones. Even Vixen was not allowed to sit near it. In the full Indian moonlight I could see a white uniform bending over the dog.

"Good-bye, old man." We could not help hearing Stanley's voice. "For 'Eving's sake don't get bit and go mad by any measly pi-dog. But you can look after yourself, old man. *You* don't get drunk an' run about 'ittin' your friends. You takes your bones an' you eats your biscuit, an' you kills your enemy like a gentleman. I'm goin' away—don't 'owl—I'm goin' off to Kasauli where I won't see you no more."

I could hear him holding Garm's nose as the dog threw it up to the stars.

"You'll stay here an' be'ave, an'—an' I'll go away an' try to be'ave, an' I don't know 'ow to leave you. I don't know—"

"I think this is damn' silly," said the officer, patting his foolish fubsy old retriever. He called to the private, who leaped to his feet, marched forward, and saluted.

"You here?" said the officer, turning away his head.

"Yes, sir, but I'm just goin' back."

"I shall be leaving here at eleven in my cart. You come with me. I can't have sick men running about all over the place. Report yourself at eleven, *here*."

We did not say much when we went indoors, but the officer muttered and pulled his retriever's ears.

He was a disgraceful, overfed door-mat of a dog; and when he waddled off to my cookhouse to be fed, I had a brilliant idea.

At eleven o'clock that officer's dog was nowhere to be found, and you never heard such a fuss as his owner made. He called and shouted and grew angry, and hunted through my garden for half an hour.

Then I said:—

"He's sure to turn up in the morning. Send a man in by rail, and I'll find the beast and return him."

"Beast?" said the officer. "I value that dog considerably more than I value any man I know. It's all very fine for you to talk. Your dog's here."

So she was—under my feet—and, had she been missing, food and wages would have stopped in my house till her return. But some people grow fond of dogs not worth a cut of the whip. My friend had to drive away at last with Stanley in the back-seat. And then the dog-boy said to me:—

"What kind of animal is Bullen Sahib's dog? Look at him!"

I went to the boy's hut, and the fat old reprobate was lying on a mat carefully chained up. He must have heard his master calling for twenty minutes, but had not even attempted to join him.

"He has no face," said the dog-boy scornfully. "He is a *punniar-kooter* [a spaniel]. He never tried to get that cloth off his jaws when his master called. Now Vixen-*baba* would have jumped through the window, and the Great Dog would have slain me with his muzzled mouth. It is true that there are many kinds of dogs."

Next evening who should turn up but Stanley. The officer had sent him back fourteen miles by rail with a note begging me to return the retriever if I had found him, and, if I had not, to offer huge rewards. The last train to camp left at half past ten, and Stanley stayed till ten talking to Garm. I argued and entreated, and even threatened to shoot the bull-terrier, but the little man was as firm as a rock, though I gave him a good dinner and talked to him most severely. Garm knew as well as I that this was the last time he could hope to see his man, and followed Stanley like a shadow. The retriever said nothing, but licked his lips after his meal and waddled off without so much as saying "Thank you" to the disgusted dog-boy.

So that last meeting was over and I felt as wretched as Garm, who moaned in his sleep all night. When we went to the office he found a place under the table close to Vixen, and dropped flat till it was time to go home. There was no more running out into the verandas, no slinking away for stolen talks with Stanley. As the weather grew warmer the dogs were forbidden to run beside the cart, but sat at my side on the seat, Vixen with her head under the crook of my left elbow, and Garm hugging the handrail.

Here Vixen was ever in great form. She had to attend to all the moving traffic, such as bullock-carts that blocked the way, and camels, and led ponies; as well as to keep up her dignity when she passed low friends running in the dust. She never yapped for yapping's sake, but her shrill, high bark was known all along the Mall, and other men's terriers ki-yied in reply, and bullock-drivers looked over their shoulders and gave us the road with a grin.

But Garm cared for none of these things. His big eyes were on the horizon and his terrible mouth was shut. There was another dog in the office who belonged to my Chief. We called him "Bob the Librarian," because he always imagined vain rats behind the bookshelves, and in hunting for them would drag out half the old newspaper-files. Bob was a well-meaning idiot, but Garm did not encourage him. He would slide his head round the door, panting, "Rats! Come along, Garm!" and Garm would shift one fore-paw over the other, and curl himself round, leaving Bob to whine at a most uninterested back. The office was nearly as cheerful as a tomb in those days.

Once, and only once, did I see Garm at all contented with his surroundings. He had gone for an unauthorized walk with Vixen early one Sunday morning, and a very young and foolish artilleryman (his battery had just moved to that part of the world) tried to steal them both. Vixen, of course, knew better than to take food from soldiers, and, besides, she had just finished her breakfast. So she trotted back with a large piece of the mutton that they issue to our troops, laid it down on my veranda, and looked up to see what I thought. I asked her where Garm was, and she ran in front of the horse to show me the way.

About a mile up the road we came across our artilleryman sitting very stiffly on the edge of a culvert with a greasy handkerchief on his knees. Garm was in front of him, looking rather pleased. When the man moved leg or hand, Garm bared his teeth in silence. A broken string hung from his collar, and the other half of it lay, all warm, in the artilleryman's still hand. He explained to me, keeping his eyes straight in front of him, that he had met this dog (he called him awful names) walking alone, and was going to take him to the Fort to be killed for a masterless pariah.

I said that Garm did not seem to me much of a pariah, but that he had better take him to the Fort if he thought best. He said he did not care to do so. I told him to go to the Fort alone. He said he did not want to go at that hour, but would follow my advice as soon as I had called off the dog. I instructed Garm to take him to the Fort, and Garm marched him solemnly up to the gate, one mile and a half under a hot sun, and I told the quarter-guard what had happened; but the young artilleryman was more angry than was at all necessary when they began to laugh. Several regiments, he was told, had tried to steal Garm in their time.

That month the hot weather shut down in earnest, and the dogs slept in the bathroom on the cool wet bricks where the bath is placed. Every morning, as soon as the man filled my bath, the two jumped in, and every morning the man filled the bath a second time. I said to him that he might as well fill a small tub specially for the dogs. "Nay," said he, smiling, "it is not their custom. They would not understand. Besides, the big bath gives them more space."

The punkah-coolies who pull the punkahs day and night came to know Garm intimately. He noticed that when the swaying fan stopped I would call out to the coolie and bid him pull with a long stroke. If the man still slept I would wake him up. He discovered, too, that it was a good thing to lie in the wave of air under the punkah. Maybe Stanley had taught him all about this in barracks. At any rate, when the punkah stopped, Garm would first growl and cock his eye at the rope, and if that did not wake the man—it nearly always did—he would tiptoe forth and talk in the sleeper's ear. Vixen was a clever little dog, but she could never connect the punkah and the coolie; so Garm gave me grateful hours of cool sleep. But he was utterly wretched —as miserable as a human being; and in his misery he clung so closely to me that other men noticed it and were envious. If I moved from one room to another Garm followed; if my pen stopped scratching, Garm's head was thrust into my hand; if I turned, half awake, on the pillow, Garm was up and at my side, for he knew that I was his only link with his master, and day and night, and night and day, his eyes asked one question —"When is this going to end?"

Living with the dog as I did, I never noticed that he was more than ordinarily upset by the hot weather, till one day at the Club a man said: "That dog of yours will die in a week or two. He's a shadow." Then I dosed

Garm with iron and quinine, which he hated; and I felt very anxious. He lost his appetite, and Vixen was allowed to eat his dinner under his eyes. Even that did not make him swallow, and we held a consultation on him, of the best man-doctor in the place; a lady-doctor, who cured the sick wives of kings; and the Deputy Inspector-General of the veterinary service of all India. They pronounced upon his symptoms, and I told them his story, and Garm lay on a sofa licking my hand.

"He's dying of a broken heart," said the lady-doctor suddenly.

" 'Pon my word," said the Deputy Inspector-General, "I believe Mrs. Macrae is perfectly right—as usual."

The best man-doctor in the place wrote a prescription, and the veterinary Deputy Inspector-General went over it afterwards to be sure that the drugs were in the proper dog-proportions; and that was the first time in his life that our doctor ever allowed his prescriptions to be edited. It was a strong tonic, and it put the old boy on his feet for a week or two; then he lost flesh again. I asked a man I knew to take him up to the Hills with him when he went, and the man came to the door with his kit packed on the top of the carriage. Garm took in the situation at one red glance. The hair rose along his back; he sat down in front of me and delivered the most awful growl I have ever heard in the jaws of a dog. I shouted to my friend to get away at once, and as soon as the carriage was out of the garden Garm laid his head on my knee and whined. So I knew his answer, and devoted myself to getting Stanley's address in the Hills.

My turn to go to the cool came late in August. We were allowed thirty days' holiday in a year, if no one fell sick, and we took it as we could be spared. My Chief and Bob the Librarian had their holiday first, and when they were gone I made a calendar, as I always did, and hung it up at the head of my cot, tearing off one day at a time till they returned. Vixen had gone up to the Hills with me five times before; and she appreciated the cold and the damp and the beautiful wood fires there as much as I did.

"Garm," I said, "we are going back to Stanley at Kasauli. Kasauli—Stanley; Stanley—Kasauli." And I repeated it twenty times. It was not Kasauli really, but another place. Still I remembered what Stanley had said in my garden on the last night, and I dared not change the name. Then Garm began to tremble; then he barked; and then he leaped up at me, frisking and wagging his tail.

"Not now," I said, holding my hand. "When I say 'Go,' we'll go, Garm." I pulled out the little blanket coat and spiked collar that Vixen always wore up in the Hills, to protect her against sudden chills and thieving leopards, and I let the two smell them and talk it over. What they said of course I do not know, but it made a new dog of Garm. His eyes were bright; and he barked joyfully when I spoke to him. He ate his food, and he killed rats for the next three weeks, and when he began to whine I had only to say "Stanley—Kasauli; Kasauli—Stanley," to wake him up. I wish I had thought of it before.

My Chief came back, all brown with living in the open air, and very angry at finding it so hot in the Plains. That same afternoon we three and Kadir Buksh began to pack for our month's holiday, Vixen rolling in and out of the bullock-trunk twenty times a minute, and Garm grinning all over and thumping on the floor with his tail. Vixen knew the routine of travelling as well as she knew my office-work. She went to the station, singing songs, on the front seat of the carriage, while Garm sat with me. She hurried into the railway carriage, saw Kadir Buksh make up my bed for the night, got her drink of water, and curled up with her black-patch eye on the tumult of the platform. Garm followed her (the crowd gave him a lane all to himself) and sat down on the pillows with his eyes blazing, and his tail a haze behind him.

We came to Umballa in the hot misty dawn, four or five men, who had been working hard for eleven months, shouting for our dâks—the two-horse travelling carriages that were to take us up to Kalka at the foot of the Hills. It was all new to Garm. He did not understand carriages where you lay at full length on your bedding, but Vixen knew and hopped into her place at once; Garm following. The Kalka Road, before the railway was built, was about forty-seven miles long, and the horses were changed every eight miles. Most of them jibbed, and kicked, and plunged, but they had to go, and they went rather better than usual for Garm's deep bay in their rear.

There was a river to be forded, and four bullocks pulled the carriage, and Vixen stuck her head out of the sliding-door and nearly fell into the water while she gave directions. Garm was silent and curious, and rather needed reassuring about Stanley and Kasauli. So we rolled, barking and yelping, into Kalka for lunch, and Garm ate enough for two.

After Kalka the road wound among the hills, and we took a curricle with half-broken ponies, which were changed every six miles. No one dreamed of a railroad to Simla in those days, for it was seven thousand feet up in the air. The road was more than fifty miles long, and the regulation pace was just as fast as the ponies could go. Here, again, Vixen led Garm from one carriage to the other, jumping into the back seat, and shouted. A cool breath from the snows met us above five miles out of Kalka, and she whined for her coat, wisely fearing a chill on the liver. I had had one made for Garm too, and, as we climbed to the fresh breezes, I put it on, and Garm chewed it uncomprehendingly, but I think he was grateful.

"Hi-yi-yi-yi!" sang Vixen as we shot round the curves; "Toot-toot-toot!" went the driver's bugle at the dangerous places, and "Yow! yow! yow!" bayed Garm. Kadir Buksh sat on the front seat and smiled. Even he was glad to get away from the heat of the Plains that stewed in the haze behind us. Now and then we would meet a man we knew going down to his work again, and he would say: "What's it like below?" and I would shout: "Hotter than cinders. What's it like up above?" and he would shout back: "Just perfect!" and away we would go.

Suddenly Kadir Buksh said, over his shoulder: "Here is Solon"; and Garm snored where he lay with his head on my knee. Solon is an unpleasant little cantonment, but it has the advantage of being cool and healthy. It is all bare and windy, and one generally stops at a rest-house near by for something to eat. I got out and took both dogs with me, while Kadir Buksh made tea. A soldier told us we should find Stanley "out there," nodding his head towards a bare, bleak hill.

When we climbed to the top we spied that very Stanley, who had given me all this trouble, sitting on a rock with his face in his hands and his overcoat hanging loose about him. I never saw anything so lonely and dejected in my life as this one little man, crumpled up and thinking, on the great grey hillside.

Here Garm left me.

He departed without a word, and, so far as I could see, without moving his legs. He flew through the air bodily, and I heard the whack of him as he flung himself at Stanley, knocking the little man clean over. They rolled on the ground together, shouting, and yelping, and hugging. I could not see which was dog and which was man, till Stanley got up and whimpered.

He told me that he had been suffering from fever at intervals, and was very weak. He looked all he said, but even while I watched, both man and dog plumped out to their natural sizes, precisely as dried apples swell in water. Garm was on his shoulder and his breast and feet all at the same time, so that Stanley spoke all through a cloud of Garm—gulping, sobbing, slavering Garm. He did not say anything that I could understand, except that he had fancied he was going to die, but that now he was quite well, and that he was not going to give up Garm any more to anybody under the rank of Beelzebub.

Then he said he felt hungry, and thirsty, and happy.

We went down to tea at the rest-house, where Stanley stuffed himself with sardines and raspberry jam, and beer, and cold mutton and pickles, when Garm wasn't climbing over him; and then Vixen and I went on.

Garm saw how it was at once. He said good-bye to me three times, giving me both paws one after another, and leaping on to my shoulder. He further escorted us, singing Hosannas at the top of his voice, a mile down the road. Then he raced back to his own master.

Vixen never opened her mouth, but when the cold twilight came, and we could see the lights of Simla across the hills, she snuffled with her nose at the breast of my ulster. I unbuttoned it, and tucked her inside. Then she gave a contented little sniff, and fell fast asleep, her head on my breast, till we bundled out at Simla, two of the four happiest people in all the world that night.

2 *Tatters and Scraps—Kennel-Terriers*

E. Œ. SOMERVILLE AND MARTIN ROSS

It is not to much to say that no one would be more shocked and surprised at being relegated to, as it were, the Heel of the Hunt, than those to whom these later thoughts refer. It is indeed almost with awe that one reflects on the greatness of soul, the self-respect, the self-consciousness (a quality that is attributed—erroneously, I believe—to man only) of a very small terrier.

Especially are these present in the terrier that is accredited to a pack of Foxhounds. When it is considered what it can accomplish with that tiny rough body, those indomitable little legs, that absurdly small head, packed with constructive intelligence yet with a brain-pan no bigger than an apple, it is impossible to deny that its high opinion of itself is justified.

There was once a little white rough dog who was—or thought he was—in charge of a small pack of Foxhounds in the South of Ireland. He was born of a noted strain of his sort in very distinguished surroundings, no less than the Kildare Hunt Establishment, and perhaps it was as well that, like Joseph, he was sold into captivity when he was too young to realize how very far he had come down in the world, geographically and otherwise. But, like Joseph, he soon rose to eminence. He was little more than a year old when his career as a Hunt terrier began, and the way in which he recognized and adapted himself to the conditions of his job came little short of genius. He soon realized the need of conserving his strength, and submitted to being parted from the Pack and carried by any available means to the Meet. Once there, however, he would break from control and, asserting himself as a foxhound among foxhounds, would become merged with the Pack.

But he had a technique of his own. Realizing that to struggle through the intricate jungles of those southern haunts of the Fox would exhaust him unduly before the more serious business began, he would, as soon as the Hounds were thrown into covert, post himself outside it on a convenient hillock, and there would sit, rigid, ears pricked, waiting for the first questioning note. When it came he would spring to attention. One could see that every fibre of his little body was hearkening for the next, and when certainty followed, Tatters—such was his unassuming name—like an arrow from the bow, would join his comrades, and in any interval of the melodious hound-music his small shrieks would fill the silence.

In the wild country which had become his there are seldom runs that do not find their earthly close in a drain or gully or similar place of refuge. Many were the occasions when the Hounds, having marked to Ground, would be found vainly endeavouring to thrust the greater, themselves, into that which could only contain the less, the Fox. The Huntsman, dragging by the stern a futile enthusiast out of the mouth of the hole, might then be seen looking over his shoulder, and asking, with pardonable violence, where that terrier had got to. "And then", as Lord Macaulay has said—certain quotations are so applicable that they must be corrupted to my purpose—"Then," I repeat, "the cry is 'Tatters!' and lo, the ranks divide, and—" but the quotation ceases to apply, for it is less like the stately stride of the Great Lord of Luna than a bullet leaving a gun when Tatters comes through the Field. He bores his way like an eel or an augur into the clamouring Pack, and if there is—as there generally is—a back-door to the drain, it is seldom long before the Fox has decided to leave by it, and, since the Hounds have been removed to a reasonable distance, the waiting and watching Field can view his retreat, relentlessly pursued by a little creature that is now dark brown and half its normal size, because drains are usually wet and muddy.

The Fox melts away—it is impossible to express the imperceptible way in which a fox can disappear. Even though he is—as an excited follower avers—the size of a chestnut horse, he is in a single instant incredibly lost to sight. The Hounds are laid on. Tatters is overtaken. Not so the Fox. In that hilly country of rocks and gullies and of what, locally, are spoken of as "His Dungeons", he, having been given his chance, far more often than not takes it.

Little Tatters soon became the idol of the Huntsman's heart. Every indulgence was his—rats, served up in a lordly trap, to gratify his zest for blood; the choicest morsels from the Hounds' kitchen; a barrel packed with straw and placed, in winter, as near the boiler-stove as possible. Even though those Kennels were not those of the Killing Kildares, Tatters was a happy little dog. A bride was found for him, white, like himself, with fascinating fair hair (her ears only, and one spot on her back) and with the similar tastes for rats and petty robbery that are so all-important in ensuring a happy married life. The little couple were established at the Huntsman's cottage, in due season a daughter blessed their union, and for a time all went well.

Unfortunately, however, Tatters began after a time to regard the comfortable cottage kitchen as his Capua, and craved for the original and sterner barrel at the Kennels. He had early developed a passion for the Dog-Hounds. He preferred their society even to that of his wife and daughter, and he lost no opportunity of slipping into their kennel and spending his days among them.

One morning, early in Tatters' third season, the Huntsman, with the Hounds, met the Master at the Kennel-gates as usual. They were going hunting. The Huntsman, a North-country little man, who had come from the Bramham Moor Hounds, saluted the Master, lifting his cap with solemnity, but said no word of greeting. Something was wrong. The Master said, not without anxiety:

"Hounds all right?"

"I can't 'ardly bear to look at 'em!" was the remarkable response.

"Good heavens! What's wrong?"

"They've killed little Tattoo!"

This was Tatters' pet name. Tears ran down the little Huntsman's red face.

" 'E got in among 'em last night, through the bars some way; I suppose they was startled like and didn't know it was 'im—"

That was a sad day's hunting. And as if to emphasize the fact of the recent tragedy, the only fox, found after a long day of trailing over the hills, got to ground immediately, and there was no little Tatters to fetch him out.

The widowed Mrs. Tatters had never showed any fancy for the Chase. Her only game was rats, and even these, without the late Tatters to back her up at critical moments, soon ceased to interest her. She developed *embonpoint* and kept the house.

Fortunately, however, their daughter, Scraps, had inherited her father's sporting tastes. At the time of his death she had just arrived at maturity. She had indeed already been occasionally taken out hunting with him, and the picture of the two little white things speeding after the pack, looking like demented grains of rice, is with me still. At the beginning of Scraps' career, before she had been established as official kennel-terrier, it was the Master's practice to take her, in company with an elderly Labrador, to a large walled-in place of seclusion, once an orchard, now given over to briars and bracken, and (which is more important) rabbits. There had Scraps been entered to her profession, and even though she knew that in the Season proper a nobler game was to be hers, yet she would collaborate with the Labrador with a very pretty enthusiasm, and on a system based on that of her father's in the field.

Regarding her comrade in the light of a Tufter, she was accustomed to send him into the briars to draw for the game, while she, on the central orchard path, struck appropriate attitudes. Her height was but little more than eight inches—or it may be more respectful to say two hands—at the withers, yet her mien at such moments had the arrogance and the splendour of the stance of a Barb stallion. Thus, tense and motionless, she would remain, her head held very high on a nobly arched neck; her ears fiercely pricked, her tail at half-cock, her attention dramatically concentrated on the Tufter's probable position in covert.

The rabbit breaks. The Labrador bursts into view in lumbering pursuit. Scraps flings herself into action. The coarse grass and bracken are so high that it is possible to track her course only by a trail of squeaks, and by occasional glimpses of what looks like a hairy lawn-tennis ball, as she bucks through the long grass, running for blood. (It is, of course, well known that in hunting through long grass the Best Dogs, or at least the Best Little Dogs, proceed in bucks without reference to their noses, like kangaroos.) In these Orchard Hunts the rabbit has never been known to fail to achieve safety considerably in advance of his pursuers; a disappointment to which, however, invariably experience has inured them.

Agreeable to both Scraps and her owner as were these unofficial hunts, it must be confessed that it was to the taste they had encouraged that the owner owed a time of prolonged and acute anxiety. Within a few hundred yards of the Kennels were great cliffs that held the Atlantic at bay. They held also, unfortunately, what the Huntsman called "a Dose of rabbits". To these cliffs, Scraps, slipping across the fields high above them, where sheep nibbled the salty grass and rabbits disported themselves, would repair whenever she could elude supervision, and there, along the face of the cliffs, invisible among bushes and rocks, she would hunt, deaf to calls and whistles, only returning at her own pleasure.

And at last there came a night when she did not return. At day-break the search for her began, and so continued in despairing depression for a long three days. Conviction was forced upon the searchers that she had either fallen from a cliff and been drowned, or had got wedged in a rabbit-hole and there had died. Search was abandoned.

It was a full four days later when

Late, late, in a gloaming, Kilmeny cam' hame.

A small, almost spectral form came creeping across the sheep-fields, and lay down, exhausted, unable to drag itself another inch, at the Kennel-door. Little Scraps had been starved free. Once before the same thing had happened to one of the Kennel-terriers, but in that unhappy case the cliffs had held their prey. Scraps' coat showed how tightly she had been jammed in the burrow. Her little body had wasted so thin that, like the dog in the fairy-tale, she might have been run through a wedding-ring. But starvation had saved her. She survived, and lived to end her days as pet and pampered Head-dog in the Master's home. There, despotically, she ruled the lessed dogs, and though the smallest of them all her authority was unquestioned.

Even after her death in the fullness of time, she still kept her underlings in order. In life she had been accustomed, when the mood took her, to leave her own comfortable basket and usurp that of a junior, a cushioned snail-shell, much coveted, though in no way superior to her own. It was not until after her death that the junior, whose name was Sheila, established undisputed possession of the desired basket. Yet on a certain night it was given to their owner to see Sheila's little snub nose poked out of the snail-shell, her eyes fixed very low down on the door. Something had come in. Sheila's little head followed the invisible visitor—invisible to the grosser human being, though not to Sheila—round the room. Her rough little head was twisted round the side of the snail-shell, always at the same low level. Then, gingerly, she slid out of the basket. Glancing over her shoulder as she went, she stole across the room and went to ground under the sofa. Scraps, the Head-dog, had reasserted herself.

3 *The Pick of the Puppies*

SIR PERCY FITZPATRICK

THERE were six puppies, and as the waggons were empty we fixed up a roomy nest in one of them for Jess and her family. There was no trouble with Jess; nobody interfered with her, and she interfered with nobody. The boys kept clear of her; but we used to take a look at her and the puppies as we walked along with the waggons; so by degrees she got to know that we would not harm them, and she no longer wanted to eat us alive if we went near and talked to her.

Five of the puppies were fat strong yellow little chaps with dark muzzles—just like their father, as Ted said; and their father was an imported dog, and was always spoken of as the best dog of the breed that had ever been in the country. I never saw him, so I do not really know what he was like—perhaps he was not a yellow dog at all; but, whatever he was, he had at that time a great reputation because he was "imported," and there were not half a dozen imported dogs in the whole of the Transvaal then. Many people used to ask what breed the puppies were—I suppose it was because poor cross faithful old Jess was not much to look at, and because no one had a very high opinion of yellow dogs in general, and nobody seemed to remember any famous yellow bull-terriers. They used to smile in a queer way when they asked the question, as if they were going to get off a joke; but when we answered "Just like their father—Buchanan's *imported* dog," the smile disappeared, and they would give a whistle of surprise and say "By Jove!" and immediately begin to examine the five yellow puppies, remark upon their ears and noses and legs, and praise them up until we were all as proud as if they had belonged to us.

Jess looked after her puppies and knew nothing about the remarks that were made, so that they did not worry her, but I often looked at the faithful old thing with her dark brindled face, cross-looking eyes and always-moving ears, and thought it jolly hard lines that nobody had a good word for her; it seemed rough on her that everyone should be glad there was only one puppy at all like the mother—the sixth one, a poor miserable little rat of a thing about half the size of the others. He was not yellow like them, nor dark brindled like Jess, but a sort of dirty pale half-and-half colour with some dark faint wavy lines all over him, as if he had tried to be brindled and failed; and he had a dark sharp wizened little muzzle that looked shrivelled up with age.

Most of the fellows said it would be a good thing to drown the odd one because he spoilt the litter and made them look as though they were not really thoroughbred, and because he was such a miserable little rat that he was not worth saving anyhow; but in the end he was allowed to live. I believe no one fancied the job of taking one of Jess's puppies away from her; moreover, as any dog was better than none, I had offered to take him rather than let him be drowned. Ted had old friends to whom he had already promised the pick of the puppies, so when I came along it was too late, and all he could promise me was that if there should be one over I might have it.

As they grew older and were able to crawl about they were taken off the waggons when we outspanned and put on the ground. Jess got to understand this at once, and she used to watch us quite quietly as we took them in our hands to put them down or lift them back again. When they were two or three weeks old a man came to the waggons who talked a great deal about dogs, and appeared to know what had to be done. He said that the puppies' tails ought to be docked, and that a bull-terrier would be no class at all with a long tail, but you should on no account clip his ears. I thought he was speaking of fox-terriers, and that with bull-terriers the position was the other way round, at that time; but as he said it was "the thing" in England, and nobody contradicted him, I shut up. We found out afterwards that he had made a mistake; but it was too late then, and Jess's puppies started life as bull-terriers up to date, with long ears and short tails.

I felt sure from the beginning that all the yellow puppies would be claimed and that I should have to take the odd one, or none at all; so I began to look upon him as mine already, and to take an interest in him and look after him. A long time ago somebody wrote that "the sense of possession turns sand into gold," and it is one of the truest things ever said. Until it seemed that this queer-looking odd puppy was going to be mine I used to think and say very much what the others did—but with this difference, that I always felt sorry for him, and sorry for Jess too, because he was like her and not like the father. I used to think that perhaps if he were given a chance he might grow up like poor old Jess herself, ugly, cross and unpopular, but brave and faithful. I felt sorry for him, too, because he was small and weak, and the other five big puppies used to push him away from his food and trample on him; and when they were old enough to play they used to pull him about by his ears and pack on to him—three or four to one—and bully him horribly. Many a time I rescued him, and many a time gave him a little preserved milk and water with bread soaked in it when the others had shouldered him out and eaten everything.

After a little while, when my chance of getting one of the good puppies seemed hopeless and I got used to the idea that I would have to take the odd one, I began to notice little things about him that no one else noticed, and got to be quite fond of the little beggar—in a kind of way. Perhaps I was turning my sand into gold, and my geese into swans; perhaps I grew fond of him simply because, finding him lonely and with no one else to depend on, I befriended him; and perhaps it was because he was always cheerful and plucky and it seemed as if there might be some good stuff in him after all. Those were the things I used to think of sometimes when feeding the little outcast. The other puppies would tumble him over and take his food from him; they would bump into him when he was stooping over the dish of milk and porridge, and his head was so big and his legs so weak that he would tip up and go heels over head into the dish. We were always picking him out of the food and scraping it off him: half the time he was wet and sticky, and the other half covered with porridge and sand baked hard by the sun.

One day just after the waggons had started, as I took a final look round the outspan place to see if anything had been forgotten, I found the little chap—who was only about four inches high—struggling to walk through the long grass. He was not big enough or strong enough to push his way—even the stems of the down-trodden grass tripped him—and he stumbled and floundered at every step, but he got up again each time with his little tail standing straight up, his head erect, and his ears cocked. He looked such a ridiculous sight that his little tragedy of "lost in the veld" was forgotten—one could only laugh.

What he thought he was doing, goodness only knows; he looked as proud and important as if he owned the whole world and knew that everyone in it was watching him. The poor little chap could not see a yard in that grass; and in any case he was not old enough to see much, or understand anything, for his eyes still had that bluish blind look that all very young puppies have, but he was marching along as full of confidence as a general at the head of his army. How he fell out of the waggon no one knew; perhaps the big puppies tumbled him out, or he may have tried to follow Jess, or have climbed over the tail-board to see what was the other side, for he was always going off exploring by himself. His little world was small, it may be—only the bedplank of the waggon and the few square yards of the ground on which they were dumped at the outspans—but he took it as seriously as any explorer who ever tackled a continent.

The others were a bit more softened towards the odd puppy when I caught up to the waggons and told them of his valiant struggle to follow; and the man who had docked the puppies' tails allowed, "I believe the rat's got pluck, whatever else is the matter with him, for he was the only one that didn't howl when I snipped them. The little cuss just gave a grunt, and turned round as if he wanted to eat me. I think he'd 'a' been terrible angry if he hadn't been so s'prised. Pity he's such an awful-looking mongrel."

But no one else said a good word for him: he was really beneath notice, and if ever they had to speak about him they called him "The Rat." There is no doubt about it he was extremely ugly, and instead of improving as he grew older, he became worse; yet, I could not help liking him and looking after him, sometimes feeling sorry for him, sometimes being tremendously amused, and sometimes—wonderful to relate—really admiring him. He was extraordinarily silent; while the others barked at nothing, howled when lonely, and yelled when frightened or hurt, the odd puppy did none of these things; in fact, he began to show many of Jess's peculiarities; he hardly ever barked, and when he did it was not a wild excited string of barks but little suppressed muffled noises, half bark and half growl, and just one or two at a time; and he did not appear to be afraid of anything, so one could not tell what he would do if he was.

One day we had an amusing instance of his nerve: one of the oxen, sniffing about the outspan, caught sight of him all alone, and filled with curiosity came up to examine him, as a hulking silly old tame ox will do. It moved towards him slowly and heavily with its ears spreadwide and its head down, giving great big sniffs at this new object, trying to make out what it was. "The Rat" stood quite still with his stumpy tail cocked up and his head a little on one side, and when the huge ox's nose was about a foot from him he gave one of those funny abrupt little barks. It was as if the object had suddenly "gone off" like a cracker, and the ox nearly tumbled over with fright; but even when the great mountain of a thing gave a clumsy plunge round and trotted off, "The Rat" was not the least frightened; he was startled, and his tail and ears flickered for a second, but stiffened up again instantly, and with another of those little barks he took a couple of steps forward and cocked his head on the other side. That was his way.

He was not a bit like the other puppies; if anyone fired off a gun or cracked one of the big whips the whole five would yell at the top of their voices and, wherever they were, would start running, scrambling and floundering as fast as they could towards the waggon without once looking back to see what they were running away from. The odd puppy would drop his bone with a start or would jump round; his ears and tail would flicker up and down for a second; then he would slowly bristle up all over, and with his head cocked first on one side and then on the other, stare hard with his half-blind bluish puppy eyes in the direction of the noise; but he never ran away.

And so, little by little, I got to like him in spite of his awful ugliness. And it really was awful! The other puppies grew big all over, but the odd one at that time seemed to grow only in one part—his tummy! The poor little chap was born small and weak; he had always been bullied and crowded out by the others, and the truth is he was half starved. The natural consequence of this was that as soon as he could walk about and pick up things for himself he made up for lost time, and filled up his middle piece to an alarming size before the other parts of his body had time to grow; at that time he looked more like a big tock-tockie beetle than a dog.

Besides the balloon-like tummy he had stick-out bandy-legs, very like a beetle's too, and a neck so thin that it made the head look enormous, and you wondered how the neck ever held it up. But what made him so supremely ridiculous was that he evidently did not know he was ugly; he walked about as if he was always thinking of his dignity, and he had that puffed-out and stuck-up air of importance that you only see in small people and bantam cocks who are always trying to appear an inch taller than they really are.

When the puppies were about a month old, and could feed on porridge or bread soaked in soup or gravy, they got to be too much for Jess, and she used to leave them for hours at a time and hide in the grass so as to have but it was making fun of me too, and I could not help knowing why; it was only necessary to put the puppies together to see the reason.

After that I stopped talking about him, and made the most of the good points he showed, and tried to discover more. It was the only consolation for having to take the leavings of the litter.

Then there came a day when something happened which might easily have turned out very differently, and there would have been no stories and no Jock to tell about; and the best dog in the world would never have been my friend and companion. The puppies had been behaving very badly, and had stolen several nekstrops and chewed up parts of one or two big whips; the drivers were grumbling about all the damage done and the extra work it gave them; and Ted, exasperated by the worry of it all, announced that the puppies were quite old enough to be taken away, and that those who had picked puppies must take them at once and look after them, or let some one else have them. When I heard him say that my heart gave a little thump from excitement, for I knew the day had come when the great question would be settled once and for all. Here was a glorious and unexpected chance; perhaps one of the others would not or could not take his, and I might get one of the good ones. . . . Of course the two big ones would be snapped up: that was certain; for, even if the men who had picked them could not take them, others had been promised puppies before me would exchange those they had already chosen for the better ones. Still, there were other chances; and I thought of very little else all day long, wondering if any of the good ones would be left; and if so, which?

In the afternoon Ted came up to where we were all lying in the shade and startled us with the momentous announcement:

"Billy Griffiths can't take his pup!"

Every man of us sat up. Billy's pup was the first pick, the champion of the litter, the biggest and strongest of the lot. Several of the others said at once that they would exchange theirs for this one; but Ted smiled and shook his head.

"No," he said, "you had a good pick in the beginning." Then he turned to me, and added: "You've only had leavings." Someone said "The Rat," and there was a shout of laughter, but Ted went on; "You can have Billy's pup."

It seemed too good to be true; not even in my wildest imaginings had I fancied myself getting the pick of the lot. I hardly waited to thank Ted before going off to look at my champion. I had seen and admired him times out of number, but it seemed as if he must look different now that he belonged to me. He was a fine big fellow, well built and strong, and looked as if he could beat all the rest put together. His legs were straight; his neck sturdy; his muzzle dark and shapely; his ears equal and well carried; and in the sunlight his yellow coat looked quite bright, with occasional glints of gold in it. He was indeed a handsome fellow.

As I put him back again with the others the odd puppy, who had stood up and sniffed at me when I came, licked my hand and twiddled his tail with the friendliest and most independent air, as if he knew me quite well and was glad to see me, and I patted the poor little chap as he waddled up. I had forgotten him in the excitement of getting Billy's pup; but the sight of him made me think of his funny ways, his pluck and independence, and of how he had not a friend in the world except Jess and me; and I felt downright sorry for him. I picked him up and talked to him; and when his wizened little face was close to mine, he opened his mouth as if laughing, and shooting out his red tongue dabbed me right on the tip of my nose in pure friendliness. The poor little fellow looked more ludicrous than ever: he had been feeding again and was as tight as a drum; his skin was so tight one could not help thinking that if he walked over a mimosa thorn and got a scratch on the tummy he would burst like a toy balloon.

I put him back with the other puppies and returned to the tree where Ted and the rest were sitting. As I came up there was a shout of laughter, and—turning round to see what had provoked it—I found "The Rat" at my heels. He had followed me and was trotting and stumbling along, tripping every yard or so, but getting up again with head erect, ears cocked and his stumpy tail twiddling away just as pleased and proud as if he thought he had really started in life and was doing what only a "really and truly" grown-up dog is supposed to do—that is, follow his master wherever he goes.

All the old chaff and jokes were fired off at me again, and I had no peace for quite a time. They all had something to say: "He won't swap you off!" "I'll back 'The Rat'!" "He is going to take care of you!" "He is afraid you'll get lost!" and so on; and they were still chaffing about it when I grabbed "The Rat" and took him back again.

Billy's failure to take his puppy was so entirely unexpected and so important that the subject kept cropping up all the evening. It was very amusing then to see how each of those who had wanted to get him succeeded in finding good reasons for thinking that his own puppy was really better than Billy's. However they differed in their estimates of each other's dogs, they all agreed that the best judge in the world could not be certain of picking out the best dog in a good litter until the puppies were several months old; and they all gave instances in which the best looking puppy had turned out the worst dog, and others in which the one has no one would look at had grown up to be the champion. Goodness knows how long this would have gone on if Robbie had not mischievously suggested that "perhaps 'The Rat' was going to beat the whole lot." There was such a chorus of guffaws at this that no one told any more stories.

The poor little friendless Rat! It was unfortunate, but the truth is that he was uglier than before; and yet I could not help liking him. I fell asleep that night thinking of the two puppies—the best and the worst in the litter. No sooner had I gone over the splendid points in Billy's pup and made up my mind that he was certainly the finest I had ever seen, than the friendly wizened little face, the half-cocked ears and head on one side, the cocky little stump of a tail, and the comical dignified plucky look of the odd puppy would all come back to me. The thought of how he had licked my hand and twiddled his tail at me, and how he dabbed me on the nose, and then the manful way in which he had struggled after me through the grass, all made my heart go soft towards him, and I fell asleep not knowing what to do.

When I woke up in the morning, my first thought was of the odd puppy—how he looked to me as his only friend, and what he would feel like if, after looking on me as really belonging to him and as the one person that he was going to take care of all his life, he knew he was to be left behind or given away to anyone who would take him. It would never have entered his head that he required someone to look after him; from the way he had followed me the night before it was clear he was looking after me; and the other fellows thought the same thing. His whole manner had plainly said: "Never mind old man! Don't you worry: I am here."

We used to make our first trek at about three o'clock in the morning, so as to be outspanned by sunrise; and walking along during that morning trek I recalled all the stories that the others had told of miserable puppies having grown into wonderful dogs, and of great men who had been very ordinary children; and at breakfast I took the plunge.

"Ted," I said, bracing myself for the laughter, "if you don't mind, I'll stick to 'The Rat.' "

4 *Rex*

D. H. LAWRENCE

SINCE every family has its black sheep, it almost follows that every man must have a sooty uncle. Lucky if he hasn't two. However, it is only with my mother's brother that we are concerned. She had loved him dearly when he was a little blond boy. When he grew up black, she was always vowing she would never speak to him again. Yet when he put in an appearance, after years of absence, she invariably received him in a festive mood, and was even flirty with him.

He rolled up one day in a dog-cart, when I was a small boy. He was large and bullet-headed and blustering, and this time, sporty. Sometimes he was rather literary, sometimes coloured with business. But this time he was in checks, and was sporty. We viewed him from a distance.

The upshot was, would we rear a pup for him. Now my mother detested animals about the house. She could not bear the mix-up of human with animal life. Yet she consented to bring up the pup.

My uncle had taken a large, vulgar public-house in a large and vulgar town. It came to pass that I must fetch the pup. Strange for me, a member of the Band of Hope, to enter the big, noisy, smelly plate-glass and mahogany public-house. It was called The Good Omen. Strange to have my uncle towering over me in the passage, shouting "Hello, Johnny, what d'yer want?" He didn't know me. Strange to think he was my mother's brother, and that he had his bouts when he read Browning aloud with emotion and éclat.

I was given tea in a narrow, uncomfortable sort of living-room, half kitchen. Curious that such a palatial pub should show such miserable private accommodations, but so it was. There was I, unhappy, and glad to escape with the soft fat pup. It was winter-time and I wore a big-flapped black overcoat, half cloak. Under the cloak-sleeves I hid the puppy, who trembled. It was Saturday, and the train was crowded, and he whimpered under my coat. I sat in mortal fear of being hauled out for travelling without a dog-ticket. However, we arrived, and my torments were for nothing.

The others were wildly excited over the puppy. He was small and fat and white, with a brown-and-black head: a fox terrier. My father said he had a lemon head—some such mysterious technical phraseology. It wasn't lemon at all, but coloured like a field bee. And he had a black spot at the root of his spine.

It was Saturday night—bath-night. He crawled on the hearth-rug like a fat white teacup, and licked the bare toes that had just been bathed.

"He ought to be called Spot," said one. But that was too ordinary. It was a great question, what to call him.

"Call him Rex—the King," said my mother, looking

down on the fat, animated little teacup, who was chewing my sister's little toe and making her squeal with joy and tickles. We took the name in all seriousness.

"Rex—the King!" We thought it was just right. Not for years did I realize that it was a sarcasm on my mother's part. She must have wasted some twenty years or more of irony on our incurable naïveté.

It wasn't a successful name, really. Because my father and all the people in the street failed completely to pronounce the monosyllable Rex. They all said Rax. And it always distressed me. It always suggested to me seaweed, and rack-and-ruin. Poor Rex!

We loved him dearly. The first night we woke to hear him weeping and whinnying in loneliness at the foot of the stairs. When it could be borne no more, I slipped down for him, and he slept under the sheets.

"I won't have that little beast in the beds. Beds are not for dogs," declared my mother callously.

"He's as good as we are!" we cried, injured.

"Whether he is or not, he's not going in the beds."

I think now, my mother scorned us for our lack of pride. We were a little *infra dig.*, we children.

The second night, however, Rex wept the same and in the same way was comforted. The third night we heard our father plod downstairs, heard several slaps administered to the yelling, dismayed puppy, and heard the amiable, but to us heartless, voice saying "Shut it then! Shut thy noise, 'st hear? Stop in thy basket, stop there!"

"It's a shame!" we shouted, in muffled rebellion, from the sheets.

"I'll give you shame, if you don't hold your noise and go to sleep," called our mother from her room. Whereupon we shed angry tears and went to sleep. But there was a tension.

"Such a houseful of idiots would make me detest the

little beast even if he was better than he is," said my mother.

But as a matter of fact, she did not detest Rexie at all. She only had to pretend to do so, to balance our adoration. And in truth, she did not care for close contact with animals. She was too fastidious. My father, however, would take on a real dog's voice, talking to the puppy: a funny, high, sing-song falsetto which he seemed to produce at the top of his head. " 'S a pretty little dog! 's a pretty little doggy!—ay!—yes!—he is, yes!—Wag thy strunt, then! Wag thy strunt, Rexie!—Ha-ha! Nay, tha munna—" This last as the puppy, wild with excitement at the strange falsetto voice, licked my father's nostrils and bit my father's nose with his sharp little teeth.

" 'E makes blood come," said my father.

"Serves you right for being so silly with him," said my mother. It was odd to see her as she watched the man, my father, crouching and talking to the little dog and laughing strangely when the little creature bit his nose and toused his beard. What does a woman think of her husband at such a moment?

My mother amused herself over the names we called him.

"He's an angel—he's a little butterfly—Rexie, my sweet!"

"Sweet! A dirty little object!" interpolated my mother. She and he had a feud from the first. Of course he chewed boots and worried our stockings and swallowed our garters. The moment we took off our stockings he would dart away with one, we after him. Then as he hung, growling, vociferously, at one end of the stocking, we at the other, we would cry:

"Look at him, mother! He'll make holes in it again." Whereupon my mother darted at him and spanked him sharply.

"Let go, sir, you destructive little fiend."

But he didn't let go. He began to growl with real rage, and hung on viciously. Mite as he was, he defied her with a manly fury. He did not hate her, nor she him. But they had one long battle with one another.

"I'll teach you, my Jockey! Do you think I'm going to spend all my life darning after your destructive little teeth! I'll show you if I will!"

But Rexie only growled more viciously. They both became really angry, whilst we children expostulated earnestly with both. He would not let her take the stocking from him.

"You should tell him properly, mother. He won't be driven," we said.

"I'll drive him further than he bargains for. I'll drive him out of my sight for ever, that I will," declared my mother, truly angry. He would put her into a real temper, with his tiny, growling definance.

"He's sweet! A Rexie, a little Rexie!"

"A filthy little nuisance! Don't think I'll put up with him."

And to tell the truth, he was dirty at first. How could he be otherwise, so young! But my mother hated him for it. And perhaps this was the real start of their hostility. For he lived in the house with us. He would wrinkle his nose and show his tiny dagger-teeth in fury when he was thwarted, and his growls of real battle-rage against my mother rejoiced us as much as they angered her. But at last she caught him *in flagrante*. She pounced on him, rubbed his nose in the mess, and flung him out into the yard. He yelped with shame and disgust and indignation. I shall never forget the sight of him as he rolled over, then tried to turn his head away from the disgust of his own muzzle, shaking his little snout with a sort of horror, and trying to sneeze it off. My sister gave a yell of despair, and dashed out with a rag and a pan of water, weeping wildly. She sat in the middle of the yard with the befouled puppy, and shedding bitter tears she wiped him and washed him clean. Loudly she reproached my mother. "Look how much bigger you are than he is. It's a shame, it's a shame!"

"You ridiculous little lunatic, you've undone all the good it would do him, with your soft ways. Why is my life made a curse with animals! Haven't I enough as it is—"

There was a subdued tension afterwards. Rex was a little white chasm between us and our parent.

He became clean. But then another tragedy loomed. He must be docked. His floating puppy-tail must be docked short. This time my father was the enemy. My mother agreed with us that it was an unnecessary cruelty. But my father was adamant. "The dog'll look a fool all his life, if he's not docked." And there was no getting away from it. To add to the horror, poor Rex's tail must be *bitten* off. Why bitten? we asked aghast. We were assured that biting was the only way. A man would take the little tail and just nip it through with his teeth, at a certain joint. My father lifted his lips and bared his incisors, to suit the description. We shuddered. But we were in the hands of fate.

Rex was carried away, and a man called Rowbotham bit off the superfluity of his tail in the Nag's Head, for a quart of best and bitter. We lamented our poor diminished puppy, but agreed to find him more manly and *comme il faut*. We should always have been ashamed of his little whip of a tail, if it had not been shortened. My father said it had made a man of him.

Perhaps it had. For now his true nature came out. And his true nature, like so much else, was dual. First he was a fierce, canine little beast, a beast of rapine and blood. He longed to hunt, savagely. He lusted to set his teeth in his prey. It was no joke with him. The old canine Adam stood first in him, the dog with fangs and glaring eyes. He flew at us when we annoyed him. He flew at all intruders, particularly the postman. He was almost a peril to the neighbourhood. But not quite. Because close second in his nature stood that fatal need to love, the *besoin d'aimer* which at last makes an end of liberty. He had a terrible, terrible necessity to love, and this trammelled the native, savage hunting beast which he was. He was torn between two great impulses: the native impulse to hunt and kill, and the strange secondary, supervening impulse to love and obey. If he had been left to my father and mother, he would have run wild and got himself shot. As it was, he loved us children with a fierce, joyous love. And we loved him.

When we came home from school we would see him standing at the end of the entry, cocking his head wistfully at the open country in front of him, and meditating whether to be off or not: a white, inquiring little figure, with green savage freedom in front of him. A cry from a far distance from one of us, and like a bullet he hurled himself down the road, in a mad game. Seeing him coming, my sister invariably turned and fled, shrieking with delighted terror. And he would leap straight up her back, and bite her and tear her clothes. But it was only an ecstasy of savage love, and she knew it. She didn't care if he tore her pinafores. But my mother did.

My mother was maddened by him. He was a little demon. At the least provocation, he flew. You had only to sweep the floor, and he bristled and sprang at the broom. Nor would he let go. With his scruff erect and his nostrils snorting rage, he would turn up the whites of his eyes at my mother, as she wrestled at the other end of the broom. "Leave go, sir, leave go!" She wrestled and stamped her foot, and he answered with horrid growls. In the end it was she who had to let go. Then she flew at him, and he flew at her. All the time we had him, he was within a hair's-breadth of savagely biting her. And she knew it. Yet he always kept sufficient self-control.

We children loved his temper. We would drag the bones from his mouth, and put him into such paroxysms of rage that he would twist his head right over and lay it on the ground upside-down because he didn't know what to do with himself, the savage was so strong in him and he must fly at us. "He'll fly at your throat one of these days," said my father. Neither he nor my mother dared have touched Rex's bone. It was enough to see him bristle and roll the whites of his eyes when they came near. How near he must have been to driving his teeth right into us, cannot be told. He was a horrid sight snarling and crouching at us. But we only laughed and rebuked him. And he would whimper in the sheer torment of his need to attack us.

He never did hurt us. He never hurt anybody, though the neighbourhood was terrified of him. But he took to hunting. To my mother's disgust, he would bring large dead bleeding rats and lay them on the hearth-rug, and she had to take them up on a shovel. For he would not remove them. Occasionally he brought a mangled rabbit, and sometimes, alas, fragmentary poultry. We were in terror of prosecution. Once he came home bloody and feathers and rather sheepish-looking. We cleaned him and questioned him and abused him. Next day we heard of six dead ducks. Thank heaven no one had seen him.

But he was disobedient. If he saw a hen he was off, and calling would not bring him back. He was worst of all with my father, who would take him walks on Sunday morning. My mother would not walk a yard with him. Once, walking with my father he rushed off at some sheep in a field. My father yelled in vain. The dog was at the sheep, and meant business. My father crawled through the hedge, and was upon him in time. And now the man was in a paroxysm of rage. He dragged the little beast into the road and thrashed him with a walking stick.

"Do you know you're thrashing that dog unmercifully?" said a passerby.

"Ay, an' mean to," shouted my father.

The curious thing was that Rex did not respect my father any the more, for the beatings he had from him. He took much more heed of us children, always.

But he let us down also. One fatal Saturday he disappeared. We hunted and called, but no Rex. We were bathed, and it was bed-time, but we would not go to bed. Instead we sat in a row in our nightdresses on the sofa, and wept without stopping. This drove our mother mad.

"Am I going to put up with it? Am I? And all for that hateful little beast of a dog! He shall go! If he's not gone now, he shall go."

Our father came in late, looking rather queer, with his hat over his eye. But in his staccato tippled fashion he tried to be consoling.

"Never mind, my duckie, I s'll look for him in the morning."

Sunday came—oh, such a Sunday. We cried, and didn't eat. We scoured the land, and for the first time realized how empty and wide the earth is, when you're looking for something. My father walked for many miles—all in vain. Sunday dinner, with rhubarb pudding, I remember, and an atmosphere of abject misery that was unbearable.

"Never," said my mother, "never shall an animal set foot in this house again, while I live. I knew what it would be! I knew."

The day wore on, and it was the black gloom of bed-time, when we heard a scratch and an impudent little whine at the door. In trotted Rex, mud black, disreputable, and impudent. His air of off-hand "How d'ye do" was indescribable. He trotted around with *suffisance,* wagging his tail as if to say, "Yes, I've come back. But I didn't need to. I can carry on remarkably well by myself." Then he walked to his water, and drank noisily and ostentatiously. It was rather a slap in the eye for us.

He disappeared once or twice in this fashion. We never knew where he went. And we began to feel that his heart was not so golden as we had imagined it.

But one fatal day reappeared my uncle and the dog-cart. He whistled to Rex, and Rex trotted up. But when he wanted to examine the lusty, sturdy dog, Rex became suddenly still, then sprang free. Quite jauntily he trotted round—but out of reach of my uncle. He leaped up, licking our faces, and trying to make us play.

"Why, what ha' you done wi' the dog—you've made a fool of him. He's softer than grease. You've ruined him. You've made a damned fool of him," shouted my uncle.

Rex was captured and hauled off to the dog-cart and tied to the seat. He was in a frenzy. He yelped and shrieked and struggled, and was hit on the head, hard, with the butt-end of my uncle's whip, which only made him struggle more frantically. So we saw him driven away, our beloved Rex, frantically, madly fighting to get to us from the high dog-cart, and being knocked down, whilst we stood in the street in mute despair.

After which, black tears, and a little wound which is still alive in our hearts.

5 *Josephine has her Day*

JAMES THURBER

THE Dickinson's pup was a failure. A bull-terrier, a female, and a failure. With all of life before her, she had suddenly gone into a decline.

"She is pining away," said Dick, "like a mid-Victorian lady whose cavalier rode off and never came back."

"No," said Ellen, "there's nothing romantic about her. She looks like a servant girl who has been caught stealing a bar pin."

The failure, which was spiritual as well as physical, was unaccountable. Three weeks before, the pup had been bright and waggly and rotund. Ellen, discovering it in a bird store yapping at the virulent green tail of an indignant lady parrot, had called it not only a little plum-plum, but also a little umpsy-dumpsy.

Thus one thing had led to another, including momentary forgetfulness of their original intention to buy a Scotch terrier, mitigation of the crime being a female and a bull-terrier, and eventual purchase of the puppy. And now here she had been shipped to their summer cottage in the Adirondacks, covered with gloom and sulphur, the shadow of her former self. They sat above her, the first hour of her arrival, in grim judgment.

"Maybe she's just growing," said Dick hopefully.

"Any idiot could see she's shrinking," said Ellen. "Of course, you would have a bull-terrier."

"I don't think it is a bull-terrier now," said Dick.

"Well it *was* a bull-terrier. And a female, too! What ever possessed us!"

"You called her a little plum-plum," murmured Dick.

"And you bought her," retorted Ellen. "Well, she'll

have to be fed, I suppose. There wasn't a thing to eat in that shipping box." She swooped up the little dog. Now that it was so thin and its excess skin so wrinkly, a curious black edging around its eyes and jaw completed an effect of the most profound melancholy.

The puppy gave only two depressed laps at the milk-soaked bread placed before her, and then wobbled over to the stove in the sitting-room, revolved uncertainly three times, and closed her eyes with a pessimistic sigh.

"Well, sir, she's a nice doggy," said Dick generously, starting over to her, "yes, sir, she's a nice doggy." But his wife intervened.

"Mustn't do that," she warned. "It says in the puppy book not to disturb them at their normal sleeping hours." Mrs. Dickinson had bought a lovely puppy book, illustrated with pictures of bright and waggly puppies.

When they went to bed the puppy was sleeping soundly on a bed they fixed for it in a corner of the kitchen, apparently glad to rest after the long, jolting ride. This gladness, however, did not carry her through the night. When the stars were still bright the Dickinsons were aroused from sleep by a clamorous yelping, a wonderfully able and lusty yelping for such a despondent puppy.

"Good Lord!" groaned Dickinson. "Now what?"

"They are bound to yelp the first few nights," said his wife, sleepily.

"Doesn't it say anything in the book about their not disturbing us at our normal sleeping hours?" demanded Dickinson. "Can't I go out and shut her up, do you suppose?"

"No. It would encourage her to expect a response every time she howled, and if you humour them that way they would soon get the upper hand."

"Well, if she keeps this up she'll get it anyway," grumbled Dick as he stuffed the ends of his pillow into his ears. "The National Association of Puppies probably hired the man to write that book."

The next morning at his typewriter Dickinson felt his mind being drawn, slowly but relentlessly, away from the necessary concentration on his work. Something was striking into his brain like the measured thud of a distant drum. "Come . . . come . . . come."

It was his wife's voice emerging from the "secluded room." Gradually a note of exasperation crept in. It was followed by the sound of a slight scratching body being dragged across the floor. There were more "comes," a silence, and more scratching. Then very insistent "comes," but no scratching.

"Is he dead, dear?" called out Dick hopefully.

His wife came into the room, carrying the pessimistic-looking puppy, still saffron with the sulphur that, Dick hazarded, had been showered upon her in the interest of "bug prevention."

"She seems listless," said Mrs. Dickinson. "Do you suppose she was the runt of a litter? The book says to avoid the runts."

"Napoleon was a runt," observed Dick sagely.

"But not of a litter," responded his wife.

"By golly," exclaimed Dick suddenly. "I got a name for her, anyway! We'll call her Josephine!"

"Josephine?"

"Yes. After the wife of Napoleon, the famous runt. She started well, but got sort of 'down and out'."

Mrs. Dickinson dropped wearily into a chair and put the puppy on the floor. "Well, I've tried to make this thing come to me all morning, and she just sits and studies the floor with that darned frown."

"Maybe she just doesn't want to come," said Dick.

"That's no reason why she shouldn't. The book says it is almost certain—wait a minute, I'll read it to you." She opened the puppy book which she carried in one hand. "Here: 'It is almost certain that you will be unexpectedly delighted during the very first lesson by their sudden scampering comprehension and that you will perceive they accept your word as law.' "

They watched the little dog study the carpet with incurious attention.

"Lawless little beast," mused Dick. "Look out if she begins to trace the design of the carpet with her paw. It's a sure sign of the end."

"No such luck," said Ellen with some bitterness.

Dick leaned down nearer the puppy. "Josephine!" he cried loudly but firmly. The puppy looked up at him as a little old lady on a train, constantly afraid of being carried past her station, might look at her care-free travelling companions. "She knows her name, anyway, and if she didn't have this awful thing on her mind she might scamper with comprehension, or whatever it is they do."

"Oh, I don't think she has any mind," cried Mrs. Dickinson, irritably. "But there the book says that calmness, toleration, and self-control are essential in training a puppy."

"That book certainly says a lot for such a little book," commented Dick.

"And for such a little puppy," said his wife scornfully, as she picked up Josephine and carried her outdoors.

As the days went on, Josephine seemed to have taken the stigma of runt literally, as a thing to be religiously lived up to. She remained undeveloped physically and, Mrs. Dickinson declared, mentally, too. She declined, with stolid indifference, in her lesson hour, to adhere to any of the rules of the puppy book. The gay enterprise of "fetch" seemed to create in her no emotion save perhaps a vague wonder as to why the bit of rolled-up paper was tossed about so often.

At length came a cold, drizzly Monday when the Dickinsons gave up. They had had the empress more than three weeks, and her favourite occupation was to sit near the stove, frowning dejectedly and quivering. Once she turned over a cold wood ash with her paw. But that was all.

"Let's give her away to some family around here," said Dick, finally. "They all have lots of kids who would be crazy to have her."

"Yes, and they all have lots of dogs," said Ellen, "big, virile, happy dogs. Besides, no one wants anything but a hound dog, a hunter, in this country."

"I still think we might find some family that would take her. Someone's dog may have died."

"Everybody has two or three dogs. They wouldn't all die."

"They might," said Dick, hopefully. "They might have been playing with the shotgun, not knowing it was loaded."

"Nobody would look at a runt, a female whatnot. They would laugh just to see us walking along with her."

"We won't walk—we'll hire the Blanchards' flivver and tour around hunting a home for her," said Dick, enlarging on his plan.

So the next morning, which dawned with a promising sun, they started off with Josephine, the condemned puppy that wouldn't grow and wouldn't learn, shivering and frowning at the wind in dismay. Every house they passed for several miles had a hound dog, or two or three big-pawed, long-eared creatures, nosing about the grounds. At last, however, when they tried their luck on a dirt byroad, they saw a small brown house hanging on a hillside, from whose environs came no mournful baying.

Dick stopped the car a little way down the road, bundled the puppy in his arms and got out. There he paused. The bright sun had been overcome by one of those rapidly driving caravans of dark clouds that ride the ranges of the north in the springtime. It began to rain. Dick put up his coat collar.

"Why shall I say we don't want her?" he asked his wife.

"Oh, just act bighearted," she laughed cheerily. "They might think you are Santa Claus."

But Dick's confidence in his own scheme melted rapidly in the rain as he carried the distressed and quivering puppy toward the front gate of the silent, weathered house. As he reached the slate-coloured mail box, which winds and rain had beaten to a dejected slant, the puppy began to behave in a singular fashion. Her insides, as Dick described it later, began to go up and down. He turned and carried Josephine quickly back to Mrs. Dickinson.

"She's dying," he said, handing her to his wife.

"Hiccups, silly," said Ellen. "She'll get over them. It's nothing."

"No decent family would want a dog with hiccups," said Dick, firmly.

So they decided to wait until the paroxysms were over. This meant an unusually long wait. Josephine proved to be an accomplished hiccuper. If an interval was so protracted as to give hope of cessation, the next hiccup was so violent as the threaten indefinite continuance. At length Dick knocked the ashes from his pipe determinedly, got out of the car and lifted Josephine down after him. He set her in the road. Then suddenly he leaped at her and barked.

The puppy plunged down into the ditch by the roadside, her ears flatly inside out in abject terror. Dick hurried after her and retrieved a very wet and very muddy Josephine.

"Have you lost your mind?" exclaimed his wife. But Dickinson held up Josephine and examined her carefully. There were no more hiccups.

"By golly," he said, "these home remedies are the goods."

He started for the house again, briskly, the dog under his arm. After a long time the door on which he gingerly knocked opened just wide enough to frame the hard, spare face of a woman.

"Well, what do you want?" she growled ominously. Josephine growled ominously, too.

"I—us—" began Dick. "That is . . . er . . . a . . . can you tell me how far it is to Dale?"

Dale was the town on the outskirts of which the Dickinsons lived, the town from which they had just come. The woman jerked her thumb.

"Two mile," she grunted.

Josephine growled. The woman slammed the door. Dick walked back to his wife.

"She said her husband won't have a dog about the place," he told her. "Seems his father was bitten by one and the boys all inherited this dread."

The kind-faced lady who came to the door of the next place lifted her hands in polite refusal. Land! she already had two dogs! "Maybe Eli Madden, the storekeeper in Dale, would take it," added the lady after a moment. "His dog was gored by a bull a week back—well, no—two weeks come Monday. You might try there."

They whirled back over the road to Madden's place, which was not far from their cottage. It happened that school was letting out and the streets were filled with children. They seemed suddenly with economy of movement to surround the Dickinsons as they got out of the car and put Josephine on the ground to stretch. She studied the insurmountable problem of dust with furrowed brow.

"Lookut the lion," yelled one boy. "Woo! woo!"

"What kind of a dog is it, mister?" asked another.

"Aw, that ain't no dog," jeered a third.

"This," said Dick, "is a very wonderful dog. We are walking around the world with it. It eats keys."

"Dick!" said his wife.

"Eats keys?" exclaimed the children in a grand chorus.

"Trunk keys, door keys, padlock keys—any kind of keys," said Dick. "It shakes them well before eating."

Mrs. Dickinson indignantly picked up Josephine and led the way into Madden's store. "Heavens!" she said to her husband as the door shut out the parting jeers of the sceptical children, "Don't ever do that again. It's bad enough to have such a dog without summoning spectators."

She walked to the counter as Eli Madden came into the store from a back room. "We understand," she said sweetly, "that you lost your dog recently and would like to have another one."

"Gored by a bull," said Madden.

"We have a splendid puppy here, an American hound terrier," pursued Mrs. Dickinson brightly. Dick coughed loudly and hurriedly.

"That is—a bull-terrier, an American bull-terrier," he said. "It's yours for the taking."

"I tell you," said Madden, picking up the dog and examining it as if it were a motor car part, "it's too young a dog for me to bother with. But Floyd Timmons might take her. Never seen a stray yit he wouldn't. If you say so I'll take her up when I go this evening. I drive right past Floyd's place."

"If you don't mind," said Mrs. Dickinson eagerly.

Just then Josephine sneezed.

"She ain't sick, is she?" asked Madden suspiciously.

"She never did that before," exclaimed Mrs. Dickinson.

"Maybe distemper," said Madden, spitting. Josephine sniffled and looked miserable.

"Well, has she got a cold?" cooed Mrs. Dickinson, picking her up. "Has she got a little cold?"

They arranged with Madden to bring the dog back when it was quite over its cold. But for a week Josephine sneezed and sniffled at frequent intervals and her nose remained very warm. Mrs. Dickinson fixed a warmer bed for her, heated the shawl on which she slept, and placed a blanket under that; she fed her meat broth every day and studied the puppy book carefully for further suggestions. On the eighth day Josephine was over her sniffling and seemed much brighter than she had ever been. She even romped a bit and tugged at an apron string that Mrs. Dickinson playfully shook in front of her. "That's a real bulldog trait," said Dick admiringly.

They took her out into the front yard and the puppy scampered about a little on the grass, and even barked a plaintive bark, at a vagrant scrap of paper. While they were watching her with amusement a man drove up to the end of the walk in a buggy.

"Have you got a dog here you don't want?" he sang out cheerfully. "My name's Timmons." The Dickinsons rose from where they were sitting on the porch.

"Oh yes," said Dickinson cordially, "yes—sure."

The man got out of his buggy and came toward the house. And immediately Josephine retreated a step toward her master and mistress and growled, a tiny funny growl.

"Quite a watchdog," said the man. "Here, pup." He stooped down and picked her up. She yielded with a wild look at Mrs. Dickinson. "You still want to part with her?"

"We really want a Scotch terrier," Dick told him. "That's the reason we are giving her up."

"I see," said the man. He shifted the dog to an easier position.

"Be sure she has a warm place to sleep," said Mrs. Dickinson, following Timmons to his buggy. "We aren't letting her sleep outdoors yet. She doesn't stand cold very well. She has had a cold and just got over it. Maybe you would want to take along the bed she has been using?"

"Oh, we got plenty of warm stuff we can bed her down in. We'll keep her in the kitchen of nights until it warms up a bit."

"She shouldn't be fed much cooked meat," went on Mrs. Dickinson. "If you could see that she got some broth now and then and some lean meat, well cut up. Milk isn't good for her, so we don't give her much of it."

Timmons tucked the dog into the robe by his side. Josephine peered with questioning eyes first at her new owner and then at her recent mistress.

"Now, I'd be willing to pay you a little something for her. You see, a bull-terrier will come in handy with the cows when my other dogs git too old, or die off."

"Not at all," said Dick.

"Oh no," said Mrs. Dickinson. "Her name is Josephine," she added.

"Oh, that's all right," said Timmons largely. He clucked to his horse and the buggy moved off. They had a glimpse of Josephine peering back from one end of the seat, and then a hand took her out of sight.

"Well, there's a big bother off your hands," said Dick cheerfully, and Mrs. Dickinson nodded assent.

Scarcely an hour of the next few days went by, however, without some remembrance of Josephine coming up between them. There was, for one thing, the puppy book, rather useless now; and the little bed in the corner to stumble over; and the stick with a piece of paper tied on one end that Mrs. Dickinson had made for a plaything; and puppy biscuits scattered about the house and grounds. After a week, however, Dickinson, absorbed in his work, had almost forgotten the dog. Then one noon at luncheon he was outlining to his wife a plot for a story. It was a tale of motored action, set in the mountain ways, with rumrunners and deputy sheriffs and a girl in a red roadster driving madly through it.

"Then," explained Dick, "as they near the old house, which they suppose abandoned, a dog suddenly barks—"

"I do hope he will feed her the right things," said Mrs. Dickinson.

Dick put the brakes to his careening motor cars.

"Who will feed what?"

"Josephine," said his wife.

"Still thinking of the empress, eh?"

"You know a man is so likely to be careless. I wish I could have spoken to his wife about her."

"She'll be all right," said Dick brightly, pushing the dish of strawberry jam nearer to Ellen. "And when we get that Scotty in New York this fall you'll forget all about her and be glad we got her a good home."

"She was mighty bright and fine that last day," mused his wife. "And she growled at him. But still, I suppose she would never grow."

"Never," agreed Dick. "She would have been cut dead by all the best dogs in New York."

"You know," said Ellen, after a time, as she began to stack up the dishes, "maybe it was because I nursed her through that sick spell. . . ." She sighed. Dick knew that sigh.

"How about running up to this chap's farm and finding out how she's getting along?" he asked. "We wouldn't have to let her see us."

"All right," said Ellen quickly. "We could just stop and ask how she is doing and I could tell his wife about the broth and lean meat."

So one afternoon they hired the Blanchards' machine again, stopped at Eli Madden's store to ask the way to Timmons' farm, and drove up the road until they sighted his name on a grey mailbox in front of a large, rambling farmhouse.

Timmons was kneeling in a small room of the barns, sorting over some implements, when Dickinson found him.

"Hello, Timmons," he said, with a worried crease in his forehead, "I just dropped in to ask about the dog."

"Darned if Norb Gibbs didn't take her," said Timmons, rising. 'He stopped up here one day and he had a little likker on him. Norb's a mean man. He's the orneriest cuss in these parts. Well, sir, your dog was runnin' round in front of the house and Norb took a fancy to her. I said I didn't want to sell, seein' as she'd bin give to us, and he kidded me, like, and said well, no, you wouldn't want to sell a gift dog—he'd just take her. And he did. Laughed when I tried to stop him."

"Can't you have the sheriff or someone get the dog?" asked Dick.

"Sheriff's 'way off to the county seat and his deputy here is a little thick with Gibbs. Nobody ever crosses him much. He's a hard man."

With a gesture of annoyance Dickinson finally asked the farmer to say nothing of the dog's disappearance. It wouldn't do for the thing to be talked about in the village. Then he went back to rejoin his wife.

"Did you see her?" she smiled wistfully.

"No," said Dick; with a great effort at lightness. "But

she's doing fine, Timmons said."

"I'm sure she is," said Mrs. Dickinson. "I've told Mrs. Timmons all about her idiosyncrasies. Well . . . I guess we must be getting back home. It looks a lot like rain."

And it did rain, a slow, depressing drizzle, as they returned, Dick hard put to it to affect an easy cheerfulness while his mind turned over and over the quandary into which Josephine—and he—had fallen. Perhaps it might be an easy matter to buy her back for Timmons. But how was he to arrange a meeting without his wife's knowing? Though his speculations ran alternately an undercurrent of exasperation at all this bother about an undesirable pup, a thin-lipped anger at the unknown brute's action, and a faint feeling of dread.

Schemes for recovering the puppy for Timmons kept formulating in his brain. He was still thinking of the problem when, next day, he walked to Madden's store to replenish his supply of tobacco. He decided to query the storekeeper about the haunts and habits of the unseen man who never left his thoughts.

"Norb Gibbs?" asked Madden. "Right there." He jerked his thumb.

Dickinson turned to observe a group of three men in one corner of the store, talking haltingly in low tones, two of them pulling at pipes, the third leaning idly against a counter.

"Oh, Norb!" called Madden.

Before Dickinson could arrange his thoughts or formulate a mode of procedure, one of the smokers turned slightly and looked at the storekeeper.

"Feller here wants to see you," continued Madden.

Dick felt his heart begin to beat rapidly and his hands at the fingertips became a little cold. The man who slouched over to him, scowling, was heavy and stockily set, with a great round face, scarred on one cheek. He was dressed in a corduroy suit, with leather boots laced to the knees. He was tremendously thick through the chest, and his wrists, where they showed under the sleeves, were scraggly black with hair. Dick stuffed his half-filled pipe into his pocket.

"You want me?" asked Gibbs, scratching his neck with big fingers of his right hand.

"Why—" began Dick. "I—yes. That is—you have a little dog I'd like to buy from you if I could."

"What dog?" demanded Gibbs.

"A puppy I believe you got from Mr. Timmons," said Dick with a wry attempt at a smile and a feeling that his voice was a little weak and that his tongue moved thickly.

"I got a pup from him," said Gibbs. "Yes." He planted his feet apart, put his pipe in one corner of his mouth, and pushed his hat a little from his forehead. "What about it?"

"Well," said Dickinson, "I gave him to Timmons and now he—that is—I—my wife and I believe we would like to have him—I mean *her*—back. How is she getting along?" He felt the question was a bit silly and out of place.

"Right well," drawled the big man. "I reckon she'll make me a good dog. Nope. Can't say I want to get shed of her."

"You wouldn't sell her?" asked Dick

"Nope."

"How about fifty dollars?" Dick hazarded hopefully. The thought went through his mind that maybe Timmons would agree to help buy back the dog. He had offered to pay something for it.

"Don't need no money," said Gibbs curtly.

"I—I want the dog very much," said Dick.

"Well, s'posin' you come get her," said Gibbs, his

voice rising. "And when you come, come big."

He turned and looked at his companions, as if inviting them to enjoy the scene. They listened silently. Madden, weighing out some nails on a scale, looked up with lifted brows.

"I got a lot o' handy cordwood around the place I use on them as I don't want prowlin' about," continued Gibbs. With a loud laugh he walked back and rejoined his friends. Madden resumed weighing his nails. Dick felt his face grow hot.

"You won't give her up, then?" he asked thickly.

"I said you come and get her," glowered Gibbs, thumbing some tobacco into his short pipe and leaning against the counter. One of his friends moved over and made more room for him.

"Maybe I will," said Dick. He was quivering slightly and his legs felt strangely strained under him.

"And maybe you'll git a clout like I've had to give yer damn dog now and again," laughed the man, brutally, showing his teeth in a grin at his companions.

Things grew a little hazy in front of Dickinson—a little hazy and red. He realized, with something like a flash of fire in his brain, that this strange brute had beaten his dog . . . Josephine. . . .

In two bounds he was across the room, for he was lithe and quick, if no match for the other in strength. Before Gibs could remove one ankle from the other, as he lounged against the counter—before he could take the pipe from his hand, Dickinson struck him full in the mouth with all the force of a long right swing.

They will talk about the fight that followed for many years to come. Gibbs, knocked to a sitting position between a bushel of potatoes and a heavy unopened barrel, was hampered by his weight and his heavy clothing, but when he got to his feet he rushed for

Dickinson like an injured bull. The swoop of the oncoming giant was powerful. Dickinson turned and, in sheer fright, ran to the door. But there he suddenly whirled. With a quick, mad, desperate movement he hurled himself straight at the feet of the charging form. Not for nothing had he dived like that at football dummies in his school days, battering his body at the swinging stuffed moleskins as a member of the scrub team—the fighting scrubs. He struck the man just above the ankles.

Gibbs went toppling clumsily over him and hit the floor with a terrific crash. He fell near a newly opened box of hammers, glistening with blue steel heads and white-labelled handles. Dick rolled over and picked himself up as the man grasped a hammer and turned on his knees. His throw was wild. The hammer crashed into an unlighted lamp high above a counter, and glass tinkled sharply as the lamp swung and creaked dismally. As Gibbs staggered to his feet, Dick jumped for a chair behind the large stove near one corner of the store. Apparently the man, another hammer in his hand, expected Dickinson to hide behind the stove, for he moved toward him with a triumphant leer on his lips.

But Dickinson did not hide. The fever of battle was on him. He darted out straight at his foe, swinging the chair up from the floor as he came. Gibbs, somewhat startled, bought his hammer stroke down squarely on the upturned legs of the chair, two of which caught him solidly under the arm. He swore and the hammer flew from his hand. His other hand went to an injured elbow as the chair dropped to the floor. He lurched for it, but Dick tackled him again and Gibbs went down on the upturned chair. Dick was behind him. With a well-directed shove of his foot he sent Gibbs into an even more ludicrous entanglement with the piece of furniture. After which he leaped upon him, hammering blows into the back of his head.

"Get up!" yelled Dick in a frenzy. "Get up, you dog stealer, you dog beater, you—!"

The man struggled to a sitting posture and rubbed blood with his sleeve from a gash under his eye. Only his unusually slow movements, his handicap of heaviness, prevented him from closing with Dick before the latter got over his frenzy of yelling, while he stood above the defenceless man, his arms flailing about him.

But now the fellow was on his knees, his heavy hands flat on the floor, and Dickinson's reason returned. He lurched for a counter and began to hurl things at the slowly moving terror. He threw boxes, cans, racks—everything he could get his hands on. Grapefruit and tomatoes began to fly. A can of peaches plumped roundly into Gibbs' chest. A seed rack bounded from his shoulders as the swishing packets clattered about the floor. The scoop of nails which the awe-struck Madden had abandoned for the floor behind the counter sang a rattling song past Gibbs' ear and spattered like shot over walls and floor. But Gibbs got to his feet and, warding off more missiles with the chair, moved forward—relentless, grim, terrible. "I'll kill ye!" he grunted in spasmodic breaths. "I'll break ye in two!"

"Come on!" howled Dick, a challenge that was half wild fear, for the counter was bare of anything else to throw. He backed rapidly for the door. His feet struck the overturned box of hammers and he sat down. Madly he reached out and picked up a hammer and threw it. It went wide. Gibbs, seeing his foe on the floor before him, towered high, flung the chair against the stove with a clanging crash and rushed. And Dick's second hammer, flung with all his remaining strength and a rasping sob in his throat, struck Norb Gibbs directly over the eye. With a look of surprise, he fell to the floor in a heap.

The next thing Dickinson knew, the store was filled with people. He was aware that a hundred questions were being asked by a hundred forms moving in and out around him. Then abruptly the crowd made way for a figure that moved hurriedly through the door.

"Sheriff Griggsby! It's the sheriff!"

The crazy thought went through Dickinson's mind that this must be a movie. Then he fainted.

When he came to, he found himself firmly held in the arm of the law. His eyes widened and a question formed in his mind. Had he killed Gibbs?

"You're going to take a ride with me," said the sheriff grimly. Dick shivered. "You and me," continued the sheriff, "are going up after that female bitch now. I like a man'll fight all hell for his dog, even if it *aint's* his."

Late one October day, when the western windows of houses were burning with orange fire, the Dickinsons stopped on a bench in Central Park and sat down. A sturdy little terrier with a sleek brown coat and very bright eyes, whose ancestry, however, would admittedly have been difficult to trace, jumped up and sat down between them.

Presently a lady went by, leading in leash a handsome and well-groomed Scotch terrier, of evident aristocracy.

"There, but for the grace of Gibbs," mused Dickinson, "goes our Scotty."

His wife patted the little terrier by her side and looked after the retreating Scotty.

"Oh," she said, "it makes a good enough dog of its kind."

6 *Dodo*

GERALD DURRELL

THE Magenpies liked the dogs, although they seized every opportunity to tease them. They were particularly fond of Roger, and he would frequently go and call on them, lying down close to the wire netting, ears pricked, while the Magenpies sat on the ground inside the cage, three inches away from his nose, and talked to him in soft, wheezy chucks, with an occasional raucous guffaw, as though they were telling him dirty jokes. They never teased Roger as much as they teased the other two, and they never attempted to lure him close to the wire with soft blandishment so that they could flap down and pull his tail, as they frequently did with both Widdle and Puke. On the whole the Magenpies approved of dogs, but they liked them to look *and* behave like dogs; so when Dodo made her appearance in our midst the Magenpies absolutely refused to believe that she was a dog, and treated her from the beginning with a sort of rowdy, jeering disdain.

Dodo was a breed known as a Dandy Dinmont. They look like long, fat, hair-covered balloons, with minute bow legs, enormous and protuberant eyes, and long flopping ears. Strangely enough it was due to Mother that this curious misshapen breed of dog made its appearance among us. A friend of ours had a pair of these beasts which had suddenly (after years of barrenness) produced a litter of six puppies. The poor man was at his wits' end trying to find good homes for all these offspring, and so Mother, good-naturedly and unthinkingly, said she would have one. She set off one afternoon to choose her puppy and, rather unwisely, selected a female. At the time it did not strike her as imprudent to introduce a bitch into a household exclusively populated by very masculine dogs. So, clasping the puppy (like a dimly conscious sausage) under one arm, Mother climbed into the car and drove home in triumph to show the new addition to the family. The puppy, determined to make the occasion a memorable one, was violently and persistently sick from the moment she got in the car to the moment she got out. The family, assembled on the veranda, viewed Mother's prize as it waddled up the path towards them, eyes bulging, minute legs working frantically to keep the long, drooping body in motion, ears flapping wildly, pausing now and then to vomit into a flower-bed.

"Oh, isn't he *sweet*?" cried Margo.

"Good God! It looks like a sea-slug," said Leslie.

"Mother! Really!" said Larry, contemplating Dodo with loathing, "where did you dig up that canine Frankenstein?"

"Oh, but he's *sweet*," repeated Margo, "What's wrong with him?"

"It's not a him, it's a her," said Mother, regarding her acquisition proudly; "she's called Dodo."

"Well, that's two things wrong with it for a start," said Larry. "It's a ghastly name for an animal, and to introduce a bitch into the house with those three lechers about is asking for trouble. Apart from that, just look at it! Look at the shape! How did it get like that? Did it have an accident, or was it borne like that?"

"Don't be silly, dear; it's the breed. They're *meant* to be like that."

"Nonsense, mother it's a monster. Who would want to deliberately produce a thing that shape?"

I pointed out that dachshunds were much the same shape, and they had been bred specially to enable them to get down holes after badgers. Probably the Dandy Dinmont had been bred for a similar reason.

"She looks as though she was bred to go down holes after sewage," said Larry.

"Don't be disgusting, dear. They're very nice little dogs, and very faithful, apparently."

"I should imagine they have to be faithful to anyone who shows interest in them: they can't possibly have many admirers in the world."

"I think you're being very nasty about her, and, anyway, you're in no position to talk about beauty; it's only skin deep after all, and before you go throwing stones you should look for the beam in *your* eye," said Margo triumphantly.

Larry looked puzzled.

"Is that a proverb, or a quotation from the *Builders' Gazette*?" he inquired.

"I think she means that it's an ill-wind that gathers no moss," said Leslie.

"You make me sick," said Margo, with dignified scorn.

"Well, join little Dodo in the flower-bed."

"Now, now," said Mother, "don't argue about it. It's my dog and I like her, so that's all that matters."

So Dodo settled in, and almost immediately showed faults in her make-up which caused us more trouble that all the other dogs put together. To begin with she had a weak hind-leg, and at any time during the day or night her hip joint was liable to come out of its socket, for no apparent reason. Dodo, who was no stoic, would greet this catastrophe with a series of piercing shrieks that worked up to a crescendo of such quivering intensity that it was unbearable. Strangely enough, her leg never seemed to worry her when she went out for walks, or gambolled with elephantine enthusiasm after a ball on the veranda. But invariably in the evening when the family were all sitting quietly, absorbed in writing or reading or knitting, Dodo's leg would suddenly leap out of its socket, she would roll on her back and utter a scream that would make everybody jump and lose control of whatever they were doing. By the time we had massaged her leg back into place Dodo would have screamed herself to exhaustion, and immediately fall into a deep and peaceful sleep, while we would be so unnerved that we would be unable to concentrate on anything for the rest of the evening.

We soon discovered that Dodo had an extremely limited intelligence. There was only room for one idea at a time in her skull, and once it was there Dodo would retain it grimly in spite of all opposition. She decided quite early in her career that Mother belonged to her, but she was not over-possessive at first until one afternoon Mother went off to town to do some shopping and left Dodo behind. Convinced that she would never see Mother again, Dodo went into mourning and waddled, howling sorrowfully, round the house, occasionally being so overcome with grief that her leg would come out of joint. She greeted Mother's return with incredulous joy, but made up her mind that from that moment she would not let Mother out of her sight, for fear she escaped again. So she attached herself to Mother with the tenacity of a limpet, never moving more than a couple of feet away at the most. If Mother sat down, Dodo would lie at her feet; if Mother had to get up and cross the room for a book or a cigarette, Dodo would accompany her, and then they would return together and sit down again, Dodo giving a deep sigh of satisfaction at the thought that once more she had foiled Mother's attempts at escape. She even insisted in being present when Mother had a bath, sitting dolefully by the tub and staring at Mother with embarrassing intensity. Any attempts to leave her outside the bathroom door resulted in Dodo howling madly and hurling herself at the door-panels, which almost invariably resulted in her hip slipping out of its socket. She seemed to be under the impression that it was not safe to let Mother go alone into the bathroom, even if she stood guard over the door. There was always the possibility, she seemed to think, that Mother might give her the slip by crawling down the plug-hole.

At first Dodo was regarded with tolerant scorn by Roger, Widdle, and Puke; they did not think much of her, for she was too fat and too low slung to walk far, and if they made any attempt to play with her it seemed to bring on an attack of persecution mania, and Dodo would gallop back to the house, howling for protection. Taken all round they were inclined to consider her a boring and useless addition to the household, until they discovered that she had one superlative and overwhelmingly delightful characteristic: she came into season with monotonous regularity. Dodo herself displayed an innocence about the facts of life that was rather touching. She seemed not only puzzled but positively scared at her sudden bursts of popularity, when her admirers arrived in such numbers that Mother had to go about armed with a massive stick. It was owing to this Victorian innocence that Dodo fell an easy victim to the lure of Puke's magnificent ginger eyebrows, and so met a fate worse than death when Mother inadvertently locked them in the drawing-room together while she supervised the making of tea. The sudden and unexpected arrival of the English padre and his wife, ushering them into the room in which the happy couple were disporting themselves, and the subsequent efforts to maintain a normal conversation, left Mother feeling limp, and with a raging headache.

To everyone's surprise (including Dodo's) a puppy was born of this union, a strange, mewling blob of a creature with its mother's figure and its father's unusual liver-and-white markings. To suddenly become a mother like that, Dodo found, was very demoralizing, and she almost had a nervous breakdown, for she was torn between the desire to stay in one spot with her puppy and the urge to keep as close to Mother as possible. We were, however, unaware of this psychological turmoil. Eventually Dodo decided to compromise, so she followed Mother around and carried the puppy in her month. She had spent a whole morning doing this before we discovered what she was up to; the unfortunate baby hung from her mouth by its head, its body swinging to and fro as Dodo waddled along at Mother's heels. Scolding and pleading having no effect, Mother was forced to confine herself to the bedroom with Dodo and her puppy, and we carried their meals up on a tray. Even this was not altogether successful, for if Mother moved out of the chair, Dodo, ever alert, would seize her puppy and sit there regarding Mother with starting eyes, ready to give chase if necessary.

"If this goes on much longer that puppy'll grow into a giraffe," observed Leslie.

"I know, poor little thing," said Mother; "but what

can I *do*? She picks it up if she sees me lighting a cigarette."

"Simplest thing would be to drown it," said Larry. "It's going to grow into the most horrifying animal, anyway. Look at its parents."

"No, indeed you won't drown it!" exclaimed Mother indignantly.

"Don't be *horrible*," said Margo; "the poor little thing."

"Well, I think it's a perfectly ridiculous situation, allowing yourself to be chained to a chair by a dog."

"It's my dog, and if I want to sit here I *shall*," said Mother firmly.

"But for how long? This might go on for months."

"I shall think of something," said Mother with dignity.

The solution to the problem that Mother eventually thought of was simple. She hired the maid's youngest daughter to carry the puppy for Dodo. The arrangement seemed to satisfy Dodo very well, and once more Mother was able to move about the house. She pottered from room to room like some Eastern potentate, Dodo pattering at her heels, and young Sophia bringing up the end of the line, tongue protruding and eyes squinting with the effort, bearing in her arms a large cushion on which reposed Dodo's strange offspring. When Mother was going to be in one spot for any length of time Sophia would place the cushion reverently on the ground, and Dodo would surge on to it and sigh deeply. As soon as Mother was ready to go to another part of the house, Dodo would get off her cushion, shake herself, and take up her position in the cavalcade, while Sophia lifted the cushion aloft as though it carried a crown. Mother would peer over her spectacles to make sure the column was ready, giving a little nod, and they would wind their way off to the next job.

Every evening Mother would go for a walk with the dogs, and the family would derive much amusement from watching her progress down the hill. Roger, as senior dog, would lead the procession, followed by Widdle and Puke. Then came Mother, wearing an enormous straw hat, which made her look like an animated mushroom, clutching in one hand a large trowel with which to dig any interesting wild plants she found. Dodo would waddle behind, ears protruding and tongue flapping, and Sophia would bring up the rear, pacing along solemnly, carrying the imperial puppy on its cushion. Mother's Circus, Larry called it, and would irritate her by bellowing out of the window:

"Oi! Lady, wot time does the big top go up, hay?"

He purchased a bottle of hair restorer for her so that, as he explained, she could conduct experiments on Sophia and try to turn her into a bearded lady.

"That's wot your show *needs,* lady," he assured her in a hoarse voice—"a bit of clarse, see? Nothing like a bearded lady for bringin' a bit o' clarse to a show."

But in spite of all this Mother continued to lead her strange caravan off into the olive-groves at five o'clock every evening.

7 *Flush*

VIRGINIA WOOLF

AFTER two such humiliations the spirit of an ordinary dog, the spirit even of an ordinary human being, might well have been broken. But Flush, for all his softness and silkiness, had eyes that blazed; had passions that leapt not merely in bright flame but sunk and smouldered. He resolved to meet his enemy face to face and alone. No third person should interrupt this final conflict. It should be fought out by the principals themselves. On the afternoon of Tuesday, the 21st of July, therefore, he slipped downstairs and waited in the hall. He had not long to wait. Soon he heard the tramp of the familiar footsteps in the street; he heard the familiar rap on the door. Mr. Browning was admitted. Vaguely aware of the impending attack and determined to meet it in the most conciliatory of spirits, Mr. Browning had come provided with a parcel of cakes. There was Flush waiting in the hall. Mr. Browning made, evidently, some well-meant attempt to caress him; perhaps he even went so far as to offer him a cake. The gesture was enough. Flush sprang upon his enemy with unparalleled violence. His teeth once more met in Mr. Browning's trousers. But unfortunately in the excitement of the moment he forgot what was most essential—silence. He barked; he flung himself on Mr. Browning, barking loudly. The sound was sufficient to alarm the household. Wilson rushed downstairs. Wilson beat him soundly. Wilson overpowered him completely. Wilson led him in ignominy away. Ignominy it was—to have attacked Mr. Browning, to have been beaten by Wilson. Mr. Browning had not lifted a finger. Taking his cakes with him, Mr. Browning proceeded unhurt, unmoved, in perfect composure, upstairs, alone to the bedroom. Flush was led away.

After two and a half hours of miserable confinement with parrots and beetles, ferns and saucepans, in the kitchen, Flush was summoned to Miss Barrett's presence. She was lying on the sofa with her sister Arabella beside her. Conscious of the rightness of his cause, Flush went straight to her. But she refused to look at him. He turned to Arabella. She merely said, "Naughty Flush, go away". Wilson was there—the formidable, the implacable Wilson. It was to her that Miss Barrett turned for information. She had beaten him, Wilson said, "because it was right". And she added, she had only beaten him with her hand. It was upon her evidence that Flush was convicted. The attack, Miss Barrett assumed had been unprovoked; she credited Mr. Browning with all virtue, with all generosity; Flush had been beaten off by a servant, without a whip, because "it was right". There was no more to be said. Miss Barrett decided against him. "So he lay down on the floor at my feet," she wrote, "looking from under his eyebrows at me." But though Flush might look, Miss Barrett refused even to meet his eyes. There she lay on the sofa; there Flush lay on the floor.

And as he lay there, exiled, on the carpet, he went through one of those whirlpools of tumultuous emotion in which the soul is either dashed upon the rocks and splintered or, finding some tuft of foothold, slowly and painfully pulls itself up, regains dry land, and at last emerges on top of a ruined universe to survey a world created afresh on a different plan. Which was it to be—destruction or reconstruction? That was the question. The outlines only of his dilemma can be traced here; for his debate was silent. Twice Flush had done his utmost to kill his enemy; twice he had failed. And why had he failed, he asked himself? Because he loved Miss Barrett. Looking up at her from under his eyebrows as she lay, severe and silent on the sofa, he knew that he must love her for ever. Things are not simple but complex. If he bit Mr. Browning he bit her too. Hatred is not hatred; hatred is also love. Here Flush shook his ears in an agony of perplexity. He turned uneasily on the floor. Mr. Browning was Miss Barrett—Miss Barrett was Mr. Browning; love is hatred and hatred is love. He stretched himself, whined and raised his head from the floor. The clock struck eight. For three hours and more he had been lying there, tossed from the horn of one dilemma to another.

Even Miss Barrett, severe, cold, implacable as she was, laid down her pen. "Wicked Flush!" she had been writing to Mr. Bowning, ". . . if people like Flush, choose to behave like dogs savagely, they must take the consequences indeed, as dogs usually do! And *you,* so good and gentle to him! Anyone but *you* would have said 'hasty words' at least." Really it would be a good plan, she thought, to buy a muzzle. And then she looked up and saw Flush. Something unusual in his look must have struck her. She paused. She laid down her pen. Once he had roused her with a kiss, and she had thought that he was Pan. He had eaten chicken and rice pudding soaked in cream. He had given up the sunshine for her sake. She called him to her and said she forgave him.

But to be forgiven, as if for a passing whim, to be taken back again on to the sofa as if he had learnt nothing in his anguish on the floor, as if he were the same dog when in fact he differed totally, was impossible. For the moment, exhausted as he was, Flush submitted. A few days later, however, a remarkable scene took place between him and Miss Barrett which showed the depths of his emotions. Mr. Browning had been and gone; Flush was alone with Miss Barrett. Normally he would have leapt on to the sofa at her feet. But now, instead of jumping up as usual and claiming her caress, Flush went to what was now called "Mr. Browning's armchair". Usually the chair was abhorrent to him; it still held the shape of his enemy. But now, such was the battle he had won, such was the charity that suffused him, that he not only looked at the chair but, as he looked, "suddenly fell into a rapture". Miss Barrett, watching him intently, observed this extraordinary portent. Next she saw him turn his eyes towards a table. On that table still lay the packet of Mr. Browning's cakes. He "reminded me that the cakes you left were on the table". They were now old cakes, stale cakes, cakes bereft of any carnal seduction. Flush's meaning was plain. He had refused to eat the cakes when they were fresh, because they were offered by an enemy. He would eat them now that they were stale, because they were offered by an enemy turned to friend, because they were symbols of hatred turned to love. Yes, he signified, he would eat them now. So Miss Barrett rose and took the cakes in her hand. And as she gave them to him she admonished him, "So I explained to him that *you* had brought them for him, and that he ought to be properly ashamed therefore for his past wickedness, and make up his mind to love you and not bite you for the future—and he was allowed to profit from your goodness to him". As he swallowed down the faded flakes of that distasteful pastry—it was mouldy, it was fly-blown, it was sour—Flush solemnly repeated, in his own language, the words she had used—he swore to love Mr. Browning and not bite him for the future.

He was instantly rewarded—not by stale cakes, not by chicken's wings, not by the caresses that were now his, nor by the permission to lie once more on the sofa at Miss Barrett's feet. He was rewarded, spiritually; yet the effects were curiously physical. Like an iron bar corroding and festering and killing all natural life beneath it, hatred had lain all these months across his soul. Now, by the cutting of sharp knives and painful surgery, the iron had been excised. Now the blood ran once more; the nerves shot and tingled; flesh formed; Nature rejoiced, as in spring. Flush heard the birds sing again; he felt the leaves growing on the trees; as he lay on the sofa at Miss Barrett's feet, glory and delight coursed through his veins. He was with them, not against them, now; their hopes, their wishes, their desires were his. Flush could have barked in sympathy with Mr. Browning now. The short, sharp words raised the hackles on his neck. "I need a week of Tuesdays," Mr. Browning cried, "then a month—a year—a life!" I, Flush echoed him, need a month—a year—a life! I need all the things that you both need. We are all three conspiritors in the most glorious causes. We are joined in sympathy. We are joined in hatred. We are joined in defiance of black and beetling tyranny. We are joined in love.—In short, all Flush's hopes now were set upon some dimly apprehended but none the less certainly emerging triumph, upon some glorious victory that was to be theirs in common, when suddenly, without a word of warning, in the midst of civilization, security and friendship—he was in a shop in Vere Street with Miss Barrett and her sister: it was the morning of Tuesday the 1st of September—Flush was tumbled head over heels into darkness. The doors of a dungeon shut upon him. He was stolen.

"This morning Arabel and I, and he with us," Miss Barrett wrote, "went in a cab to Vere Street where we had a little business, and he followed us as usual into a shop and out of it again, and was at my heels when I stepped up into the carriage. Having turned, I said 'Flush', and Arabel looked round for Flush—there was no Flush! He had been caught up in that moment, from *under* the wheels, do you understand?" Mr. Browning understood perfectly well. Miss Barrett had forgotten the chain; therefore Flush was stolen. Such, in the year 1846, was the law of Wimpole Street and its neighbourhood.

Nothing, it is true, could exceed the apparent solidity and security of Wimpole Street itself. As far as an invalid could walk or a bath-chair could trundle nothing met the eye but an agreeable prospect of four-storeyed houses, plate-glass windows and mahogany doors. Even a carriage and pair, in the course of an afternoon's airing, need not, if the coachman were discreet, leave the limits of decorum and respectability. But if you were not an invalid, if you did not possess a carriage and pair, if you were—and many people were—active and able-bodied and fond of walking, then you might see sights and hear language and smell smells, not a stone's throw from Wimpole Street, that threw doubts upon the solidity even of Wimpole Street itself. So Mr. Thomas Beames found when about this time he took it into his head to go walking about London. He was surprised;

indeed he was shocked. Splendid buildings raised themselves in Westminster, yet just behind them were ruined sheds in which human beings lived herded together above herds of cows—"two in each seven feet of space". He felt that he ought to tell people what he had seen. Yet how could one describe politely a bedroom in which two or three families lived above a cow-shed, when the cow-shed had no ventilation, when the cows were milked and killed and eaten under the bedroom? That was a task, as Mr. Beames found when he came to attempt it, that taxed all the resources of the English language. And yet he felt that he ought to describe what he had seen in the course of an afternoon's walk through some of the most aristocratic parishes in London. The risk of typhus was so great. The rich could not know what dangers they were running. He could not altogether hold his tongue when he found what he did find in Westminster and Paddington and Marylebone. For instance, here was an old mansion formerly belonging to some great nobleman. Relics of marble mantelpieces remained. The rooms were panelled and the banisters were carved, and yet the floors were rotten, the walls dripped with filth; hordes of half-naked men and women had taken up their lodging in the old banqueting-halls. Then he walked on. Here an enterprising builder had pulled down the old family mansion. He had run up a jerry-built tenement house in its place. The rain dripped through the roof and the wind blew through the walls. He saw a child dipping a can into a bright-green stream and asked if they drank the water. Yes, and washed in it too, for the landlord only allowed water to be turned on twice a week. Such sights were the more surprising, because one might come upon them in the most sedate and civilized quarters of London—"the most aristocratic parishes have their share". Behind Miss Barrett's bedroom, for instance, was one of the worst slums in London. Mixed up with that respectability was this squalor. But there were certain quarters, of course, which had long been given over to the poor and were left undisturbed. In Whitechapel, or in a triangular space of ground at the bottom of the Tottenham Court Road, poverty and vice and misery had bred and seethed and propagated their kind for centuries without interference. A dense mass of aged buildings in St. Giles's was "wellnigh a penal settlement, a pauper metropolis in itself". Aptly enough, where the poor conglomerated thus, the settlement was called a Rookery. For there human beings swarmed on top of each other as rooks swarm and blacken tree-tops. Only the buildings here were not trees; they were hardly any longer buildings. They were cells of brick intersected by lanes which ran with filth. All day the lanes buzzed with half-dressed human beings; at night there poured back again into the stream the thieves, beggars and prostitutes who had been plying their trade all day in the West End. The police could do nothing. No single wayfarer could do anything except hurry through as fast as he could and perhaps drop a hint, as Mr. Beames did, with many quotations, evasions and euphemisms, that all was not quite as it should be. Cholera would come, and perhaps the hint that cholera would give would not be quite so evasive.

But in the summer of 1846 that hint had not yet been given; and the only safe course for those who lived in Wimpole Street and its neighbourhood was to keep strictly within the respectable area and to lead your dog on a chain. If one forgot, as Miss Barrett forgot, one paid the penalty, as Miss Barrett was now to pay it. The terms upon which Wimpole Street lived cheek by jowl with St. Giles's were well known. St. Giles's stole what St. Giles's could; Wimpole Street paid what Wimpole Street must. Thus Arabel at once "began to comfort me by showing how certain it was that I should recover him for ten pounds at most". Ten pounds, it was reckoned, was about the price that Mr. Taylor would ask for a cocker spaniel. Mr. Taylor was the head of the gang. As soon as a lady in Wimpole Street lost her dog she went to Mr. Taylor; he named his price, and it was paid; or if not, a brown paper parcel was delivered in Wimpole Street a few days later containing the head and paws of the dog. Such, at least, had been the experience of a lady in the neighbourhood who had tried to make terms with Mr. Taylor. But Miss Barrett of course intended to pay. Therefore when she got home she told her brother Henry, and Henry went to see Mr. Taylor that afternoon. He found him "smoking a cigar in a room with pictures"—Mr. Taylor was said to make an income of two or three thousand a year out of the dogs of Wimpole Street—and Mr. Taylor promised that he would confer with his "Society" and that the dog would be returned next day. Vexatious as it was, and especially annoying at a moment when Miss Barrett needed all her money, such were the inevitable consequences of forgetting in 1846 to keep one's dog on a chain.

But for Flush things were very different. Flush, Miss Barrett reflected, "doesn't know that we can recover him"; Flush had never mastered the principles of human society. "All this night he will howl and lament, I know perfectly", Miss Barrett wrote to Mr. Browning on the afternoon of Tuesday, the 1st September. But while Miss Barrett wrote to Mr. Browning, Flush was going through the most terrible experience of his life. He was bewildered in the extreme. One moment he was in Vere Street, among ribbons and laces; the next he was tumbled head over heels into a bag; jolted rapidly across streets, and at length was tumbled out—here. He found himself in complete darkness. He found himself in chillness and dampness. As his giddiness left him he made out a few shapes in a low dark room—broken chairs, a tumbled mattress. Then he was seized and tied tightly by the leg to some obstacle. Something sprawled on the floor—whether beast or human being, he could not tell. Great boots and draggled skirts kept stumbling in and out. Flies buzzed on scraps of old meat that were decaying on the floor. Children crawled out from dark corners and pinched his ears. He whined, and a heavy hand beat him over the head. He cowered down on the few inches of damp brick against the wall. Now he could see that the floor was crowded with animals of different kinds. Dogs tore and worried a festering bone that they had got between them. Their ribs stood out from their coats—they were half famished, dirty, diseased, uncombed, unbrushed; yet all of them, Flush could see, were dogs of the highest breeding, chained dogs, footmen's dogs, like himself.

He lay, not daring even to whimper, hour after hour. Thirst was his worst suffering; but one sip of the thick greenish water that stood in a pail near him disgusted him; he would rather die than drink another. Yet a majestic greyhound was drinking greedily. Whenever the door was kicked open he looked up. Miss Barrett—was it Miss Barrett? Had she come at last? But it was only a hairy ruffian, who kicked them all aside and stumbled to a broken chair upon which he flung himself. Then gradually the darkness thickened. He could scarcely make out what shapes those were on the floor, on the mattress, on the broken chairs. A stump of candle was stuck on the ledge over the fireplace. A flare burnt in the gutter outside. By its flickering, coarse light Flush could see terrible faces passing outside, leering at the window. Then in they came, until the small crowded room became so crowded that he had to shrink back and lie even closer against the wall. These horrible monsters—some were ragged, others were flaring with paint and feathers—squatted on the floor; hunched themselves over the table. They began to drink; they cursed and struck each other. Out tumbled from the bags that were dropped on the floor, more dogs—lap-dogs, setters, pointers, with their collars still on them; and a giant cockatoo that flustered and fluttered its way from corner to corner, shrieking "Pretty Poll", "Pretty Poll", with an accent that would have terrified its mistress, a widow in Maida Vale. Then the women's bags were opened, and out were tossed on to the table bracelets and rings and brooches such as Flush had seen Miss Barrett wear and Miss Henrietta. The demons pawed and clawed them; cursed and quarrelled over them, The dogs barked. The children shrieked, and the splendid cockatoo—such a bird as Flush had often seen pendant in a Wimpole Street window—shrieked "Pretty Poll! Pretty Poll!" faster and faster until a slipper was thrown at it and it flapped its great yellow-stained dove-grey wings in frenzy. Then the candle toppled over and fell. The room was dark. It grew steadily hotter and hotter; the smell, the heat were unbearable, Flush's nose burnt; his coat twitched. And still Miss Barrett did not come.

Miss Barrett lay on her sofa in Wimpole Street. She was vexed; she was worried, but she was not seriously alarmed. Of course Flush would suffer; he would whine and bark all night; but it was only a question of a few hours. Mr. Taylor would name his sum; she would pay it; Flush would be returned.

The morning of Wednesday the 2nd September dawned in the rookeries of Whitechapel. The broken windows gradually became smeared with grey. Light fell upon the hairy faces of ruffians lying sprawled upon the floor. Flush woke from a trance that had veiled his eyes and once more realized the truth. This was now the truth—this room, these ruffians, these whining, snapping, tightly tethered dogs, this murk, this dampness. Could it be true that he had been in a shop, with ladies, among ribbons, only yesterday? Was there such a place as Wimpole Street? Was there a room where fresh water sparkled in a purple jar; had he lain on cushions; had he been given a chicken's wing nicely roasted; and had he been torn with rage and jealousy and bitten a man with yellow gloves? The whole of that life and its emotions floated away, dissolved, became unreal.

Here, as the dusty light filtered in, a woman heaved herself off a sack and staggered out to fetch beer. The drinking and the cursing began again. A fat woman held him up by his ears and pinched his ribs, and some odious joke was made about him—there was a roar of laughter as she threw him on the floor again. The door was kicked open and banged to. Whenever that happened he looked up. Was it Wilson? Could is possibly be Mr. Browning? Or Miss Barrett? But no—it was only another thief, another murderer; he cowered back at the mere sight of those draggled skirts, of those hard, horny boots. Once he tried to gnaw a bone that was hurled his way. But his teeth could not meet in stony flesh and the rank smell disgusted him. His thirst increased and he was forced to lap a little of the green water that had been spilt from the pail. But as Wednesday wore on and he became hotter and more parched and still more sore, lying on the broken boards, one thing merged in another. He scarcely noticed what was happening. It was only when the door opened that he raised his head and looked. No, it was not Miss Barrett.

Miss Barrett, lying on the sofa in Wimpole Street, was becoming anxious. There was some hitch in the proceedings. Taylor had promised that he would go down to Whitechapel on Wednesday afternoon and confer with "the Society". Yet Wednesday afternoon, Wednesday evening passed and still Taylor did not come. This could only mean, she supposed, that the price was going to be raised—which was inconvenient enough at the moment. Still, of course, she would have to pay it. "I must have my Flush, you know", she wrote to Mr. Browning. "I can't run any risk and bargain and haggle." So she lay on the sofa writing to Mr. Browning and listening for a knock at the door. But Wilson came up with the letters; Wilson came up with the hot water. It was time for bed and Flush had not come.

Thursday the 3rd of September dawned in Whitechapel. The door opened and shut. The red setter who had been whining all night beside Flush on the floor was hauled off by a ruffian in a moleskin vest—to what fate? Was it better to be killed or to stay here? Which was worse—this life or that death? The racket, the hunger and the thirst, the reeking smells of the place—and once, Flush remembered, he had detested the scent of eau-de-Cologne—were fast obliterating any clear image, any single desire. Fragments of old memories began turning in his head. Was that the voice of old Dr. Mitford shouting in the field? Was that Kerenhappoch gossiping with the baker at the door? There was a rattling in the room and he thought he heard Miss Mitford tying up a bunch of geraniums. But it was only the wind—for it was stormy today—battering at the brown paper in the broken window pane. It was only some drunken voice raving in the gutter. It was only the old hag in the corner mumbling on and on and on as she fried a herring in a pan over a fire. He had been forgotten and deserted. No help was coming. No voice spoke to him—the parrots cried "Pretty Poll, Pretty Poll", and the canaries kept up their senseless cheeping and chirping.

Then again evening darkened the room; the candle was stuck in its saucer; the coarse light flared outside;

hordes of sinister men with bags on their backs, or garish women with painted faces, began to shuffle in at the door and to fling themselves down on the broken beds and tables. Another night had folded its blackness over Whitechapel. And the rain dripped steadily through a hole in the roof and drummed into a pail that had been stood to catch it. Miss Barrett had not come.

Thursday dawned in Wimpole Street. There was no sign of Flush—no message from Taylor. Miss Barrett was very much alarmed. She made enquiries. She summoned her brother Henry, and cross-examined him. She found out that he had tricked her. "The archfiend" Taylor had come according to his promise the night before. He had stated his terms—six guineas for the Society and half a guinea for himself. But Henry, instead of telling her, had told Mr. Barrett, with the result, of course, that Mr. Barrett had ordered him not to pay, and to conceal the visit from his sister. Miss Barrett was "very vexed and angry". She bade her brother to go at once to Mr. Taylor and pay the money. Henry refused and "talked of Papa". But it was no use talking of Papa, she protested. While they talked of Papa, Flush would be killed. She made up her mind. If Henry would not go, she would go herself: ". . . if people won't do as I choose, I shall go down tomorrow morning, and bring Flush back with me", she wrote to Mr. Browning.

But Miss Barrett now found that it was easier to say this than to do it. It was almost as difficult for her to go to Flush as for Flush to come to her. All Wimpole Street was against her. The news that Flush was stolen and that Taylor demanded a ransom was now public property. Wimpole Street was determined to make a stand against Whitechapel. Blind Mr. Boyd sent word that in his opinion it would be "an awful sin" to pay the ransom. Her father and her brother were in league against her and were capable of any treachery in the interests of their class. But worst of all—far worse—Mr. Browning himself threw all his weight, all his eloquence, all his learning, all his logic, on the side of Wimpole Street and against Flush. If Miss Barrett gave way to Taylor, he wrote, she was giving way to tyranny; she was giving way to blackmailers; she was increasing the power of evil over right, of wickedness over innocence. If she gave Taylor his demand, ". . . how will the poor owners fare who have not money enough for their dogs' redemption?" His imagination took fire; he imagined what he would say if Taylor asked him even for five shillings; he would say, "*You* are responsible for the proceedings of your gang, and *you* I mark—don't talk nonsense to me about cutting off heads or paws. Be as sure as that I stand here and tell you, I will spend my whole life in putting you down, the nuisance you declare yourself—and by every imaginable means I will be the death of you and as many of your accomplices as I can discover—but *you* I have discovered and will never lose sight of. . . ." So Mr. Browning would have replied to Taylor if he had had the good fortune to meet that gentleman. For indeed, he went on, catching a later post with a second letter that same Thursday afternoon, ". . . it is horrible to fancy how all the oppressors in their several ranks may, if they choose, twitch back to them by the heartstrings after various modes the weak and silent whose secret they have found out". He did not blame Miss Barrett—nothing she did could be anything but perfectly right, perfectly acceptable to him. Still, he continued on Friday morning. "I think it lamentable weakness . . ." If she encouraged Taylor who stole dogs, she encouraged Mr. Barnard Gregory who stole characters. Indirectly, she was responsible for all the wretches who cut their throats or fly the country because some blackmailer like Barnard Gregory took down a directory and blasted their characters. "But why write all this string of truisms about the plainest thing in the world?" So Mr. Browning stormed and vociferated from New Cross twice daily.

Lying on her sofa, Miss Barrett read the letters. How easy it would have been to yield—how easy it would have been to say, "Your good opinion is worth more to me than a hundred cocker spaniels". How easy it would have been to sink back on her pillows and sigh, "I am a weak woman; I know nothing of law and justice; decide for me". She had only to refuse to pay the ransom; she had only to defy Taylor and his Society. And if Flush were killed, if the dreadful parcel came and she opened it and out dropped his head and paws, there was Robert Browning by her side to assure her that she had done right and earned his respect. But Miss Barrett was not to be intimidated. Miss Barrett took up her pen and refuted Robert Browning. It was all very well, she said, to quote Donne; to cite the case of Gregory; to invent spirited replies to Mr. Taylor—she would have done the same had Taylor struck her; had Gregory defamed her; —would that they had! But what would Mr. Browning have done if the banditti had stolen her; had *her* in their power; threatened to cut off *her* ears and send them by post to New Cross? Whatever he would have done, her mind was made up. Flush was helpless. Her duty was to him. "But Flush, poor Flush, who has loved me so faithfully; have I a right to sacrifice *him* in his innocence, for the sake of any Mr. Taylor's guilt in the world?" Whatever Mr. Bowning might say, she was going to rescue Flush, even if she went down into the jaws of Whitechapel to fetch him, even if Robert Browning despised her for doing so.

On Saturday, therefore, with Mr. Browning's letter lying open on the table before her, she began to dress. She read his "one word more—in all this, I labour against the execrable policy of the world's husbands, fathers, brothers and domineerers in general". So, if she went to Whitechapel she was siding against Robert Browning, and in favour of fathers, brothers and domineers in general. Still, she went on dressing. A dog howled in the mews. It was tied up, helpless in the power of cruel men. It seemed to her to cry as it howled: "Think of Flush". She put on her shoes, her cloak, her hat. She glanced at Mr. Browning's letter once more. "I am about to marry you", she read. Still the dog howled. She left her room and went downstairs.

Henry Barrett met her and told her that in his opinion she might well be robbed and murdered if she did what she threatened. She told Wilson to call a cab. All trembling but submissive, Wilson obeyed. The cab came. Miss Barrett told Wilson to get in. Wilson, though convinced that death awaited her, got in. Miss Barrett told the cabman to drive to Manning Street, Shoreditch. Miss Barrett got in herself and off they drove. Soon they were beyond plate-glass windows, the mahogany doors and the area railings. They were in a world that Miss Barrett had never seen, had never guessed at. They were in a world where cows are herded under the bedroom floors, where whole families sleep in rooms with broken windows; in a world where water is turned on only twice a week, in a world where vice and poverty breed vice and poverty. They had come to a region unknown to respectable cab-drivers. The cab stopped; the driver asked his way at a public-house. "Out came two or three men. 'Oh, you want to find Mr. Taylor, I daresay!' " In this mysterious world a cab with two ladies could only come upon one errand, and that errand was already known. It was sinister in the extreme. One of the men ran into a house, and came out saying that Mr. Taylor " 'wasn't at home! but wouldn't I get out?' Wilson, in an aside of terror, entreated me not to think of such a thing." A gang of men and boys pressed round the cab. "Then wouldn't I see Mrs. Taylor?" the man asked. Miss Barrett had no wish whatever to see Mrs. Taylor; but now an immense fat woman came out of the house, "fat enough to have had an easy conscience all her life", and informed Miss Barrett that her husband was out: "might be in in a few minutes, or in so many hours—wouldn't I like to get out and wait?" Wilson tugged at her gown. Imagine waiting in the house of that woman! It was bad enough to sit in the cab with the gang of men and boys pressing round them. So Miss Barrett parleyed with the "immense feminine bandit" from the cab. Mr. Taylor had her dog, she said; Mr. Taylor had promised to restore her dog; would Mr. Taylor bring back her dog to Wimpole Street for certain that very day? "Oh yes, certainly," said the fat woman with the most gracious of smiles. She did believe that Taylor had left home precisely on that business. And she "poised her head to right and left with the most easy grace".

So the cab turned round and left Manning Street, Shoreditch. Wilson was of opinion that "we had escaped with our lives barely". Miss Barrett herself had been alarmed. "Plain enough it was that the gang was strong there. The Society, the 'Fancy' . . . had their roots in the ground", she wrote. Her mind teemed with thoughts, her eyes were full of pictures. This, then, was what lay on the other side of Wimpole Street—these faces, these houses. She had seen more while she sat in the cab at the public-house than she had seen during the five years that she had lain in the back bedroom at Wimpole Street. "The faces of those men!" she exclaimed. They were branded on her eyeballs. They stimulated her imagination as "the divine marble presences", the busts on the bookcase, had never stimulated it. Here lived women like herself; while she lay on her sofa, reading, writing, they lived thus. But the cab was now trundling along between four-storeyed houses again. Here was the familiar avenue of doors and windows: the pointed brick, the brass knockers, the regular curtains. Here was Wimpole Street and No. 50. Wilson sprang out—with what relief to find herself in safety can be imagined. But Miss Barrett perhaps hesitated a moment. She still saw "the faces of those men". They were to come before her again years later when she sat writing on a sunny balcony in Italy. They were to inspire the most vivid passages in *Aurora Leigh.* But now the butler had opened the door, and she went upstairs to her room again.

Saturday was the fifth day of Flush's imprisonment. Almost exhausted, almost hopeless, he lay panting in his dark corner of the teeming floor. Doors slammed and banged. Rough voices cried. Women screamed. Parrots chattered as they had never chattered to widows in Maida Vale, but now evil old women merely cursed at them. Insects crawled in his fur, but he was too weak, too indifferent to shake his coat. All Flush's past life and its many scenes—Reading, the greenhouse, Miss Mitford, Mr. Kenyon, the bookcases, the busts, the peasants on the blind—had faded like snowflakes dissolved in a cauldron. If he still held to hope, it was to something nameless and formless; the featureless face of someone he still called "Miss Barrett". She still existed; all the rest of the world was gone; but she still existed, though such gulfs lay between them that it was impossible, almost, that she should reach him still. Darkness began to fall again, such darkness as seemed almost able to crush out his last hope—Miss Barrett.

In truth, the forces of Wimpole Street were still, even at this last moment, battling to keep Flush and Miss Barrett apart. On Saturday afternoon she lay and waited for Taylor to come, as the immensely fat woman had promised. At last he came, but he had not brought the dog. He sent up a message—Let Miss Barrett pay him six guineas on the spot, and he would go straight to Whitechapel and fetch the dog "on his word of honour". What "the archfiend" Taylor's word of honour might be worth, Miss Barrett could not say; but "there seemed no other way for it"; Flush's life was at stake; she counted out the guineas and sent them down to Taylor in the passage. But as ill luck would have it, as Taylor waited in the passage among the umbrellas, the engravings, the pile carpet and other valuable objects, Alfred Barrett came in. The sight of the archfiend actually in the house made him lose his temper. He burst into a rage. He called him "a swindler, and a liar and a thief". Thereupon Mr. Taylor cursed him back. What was far worse, he swore that "as he hoped to be saved, we should never see our dog again", and rushed out of the house. Next morning, then, the blood-stained parcel would arrive.

Miss Barrett flung on her clothes again and rushed downstairs. Where was Wilson? Let her call a cab. She was going back to Shoreditch instantly. Her family came running to prevent her. It was getting dark. She was exhausted already. The adventure was risky enough for a man in health. For her it was madness. So they told her. Her brothers, her sisters, all came round her threatening her, dissuading her, "crying out against me for being 'quite mad' and obstinate and wilful—I was called as many names as Mr. Taylor". But she stood her ground. At last they realized the extent of her folly. Whatever the risk might be they must give way to her. Septimus promised if Ba would return to her room "and be in good humour" he would go to Taylor's himself and pay the money and bring back the dog.

So the dusk of the 5th of September faded into the blackness of night in Whitechapel. The door of the room was once more kicked open. A hairy man hauled Flush by the scruff of his neck out of his corner. Looking up into the hideous face of his old enemy, Flush did not know whether he was being taken to be killed or to be freed. Save for one phantom memory, he did not care. The man stooped. What were those great fingers fumbling at his throat for? Was it a knife or a chain? Stumbling, half blinded, on legs that staggered, Flush was led out into the open air.

In Wimpole Street Miss Barrett could not eat her dinner. Was Flush dead, or was Flush alive? She did not know. At eight o'clock there was a rap on the door; it was the usual letter from Mr. Browning. But as the door opened to admit the letter, something rushed in also;—Flush. He made straight for his purple jar. It was filled three times over; and still he drank. Miss Barrett watched the dazed, bewildered dirty dog, drinking. "He was not so enthusiastic about seeing me as I expected", she remarked. No, there was only one thing in the world he wanted—clean water.

After all, Miss Barrett had but glanced at the faces of those men and she remembered them all her life. Flush had been at their mercy in their midst for five whole days. Now as he lay on cushions once more, cold water was the only thing that seemed to have any substance, any reality. He drank continually. The old gods of the bedroom—the bookcase, the wardrobe, the busts—seemed to have lost their substance. This room was no longer the whole world; it was only a shelter. It was only a dell arched over by one trembling dock-leaf in a forest where wild beasts prowled and venomous snakes coiled; where behind every tree lurked a murderer ready to pounce. As he lay dazed and exhausted on the sofa at Miss Barrett's feet the howls of tethered dogs, the screams of birds in terror still sounded in his ears. When the door opened he started, expecting a hairy man with a knife—it was only Mr. Kenyon with a book; it was only Mr. Browning with his yellow gloves. But he shrank away from Mr. Kenyon and Mr. Browning now. He trusted them no longer. Behind those smiling, friendly faces was treachery and cruelty and deceit. Their caresses were hollow. He dreaded even walking with Wilson to the pillar-box. He would not stir without his chain. When they said, "Poor Flush, did the naughty men take you away?" he put up his head and moaned and yelled. A whip cracking sent him bolting down the area-steps into safety. Indoors he crept closer to Miss Barrett on the sofa. She alone had not deserted him. He still kept some faith in her. Gradually some substance returned to her. Exhausted, trembling, dirty and very thin, he lay on the sofa at her feet.

As the days passed and the memory of Whitechapel grew fainter, Flush, lying close to Miss Barrett on the sofa, read her feelings more clearly than ever before. They had been parted; now they were together. Indeed they had never been so much akin. Every start she gave, every movement she made, passed through him too. And she seemed now to be perpetually starting and moving. The delivery of a parcel even made her jump. She opened the parcel; with trembling fingers she took out a pair of thick boots. She hid them instantly in the corner of the cupboard. Then she lay down as if nothing had happened; yet something had happened. When they were alone she rose and took a diamond necklace from a drawer. She took out the box that held Mr. Browning's letters. She laid the boots, the necklace and the letters all in a carpet-box together and then—as if she heard a step on the stair—she pushed the box under the bed and lay down hastily, covering herself with her shawl again. Such signs of secrecy and stealth must herald, Flush felt, some approaching crisis. Were they about to fly together? Were they about to escape together from this awful world of dog-stealers and tyrants? Oh that it were possible! He trembled and whined with excitement; but in her low voice Miss Barrett bade him be quiet, and instantly he was quiet. She was very quiet too. She lay perfectly still on the sofa directly any of her brothers or sisters came in; she lay and talked to Mr. Barrett as she always lay and talked to Mr. Barrett.

But on Saturday, the 12th of September, Miss Barrett did what Flush had never known her to do before. She dressed herself as if to go out directly after breakfast. Moreover, as he watched her dress, Flush knew perfectly well from the expression on her face that he was not to go with her. She was bound on secret business of her own. At ten Wilson came into the room. She also was dressed as if for a walk. They went out together; and Flush lay on the sofa and waited for their return. An hour or so later Miss Barrett came back alone. She did not look at him—she seemed to look at nothing. She drew off her gloves and for a moment he saw a gold band shine on one of the fingers of her left hand. Then he saw her slip the ring from her hand and hide it in the darkness of a drawer. Then she laid herself down as usual on the sofa. He lay by her side scarcely daring to breathe, for whatever had happened, it was something that must at all costs be concealed.

At all costs the life of the bedroom must go on as usual. Yet everything was different. The very movement of the blind as it drew in and out seemed to Flush like a signal. And as the lights and shadows passed over the busts they too seemed to be hinting and beckoning. Everything in the room seemed to be aware of change; to be prepared for some event. And yet all was silent; all was concealed. The brothers and sisters came in and out as usual; Mr. Barrett came as usual in the evening. He looked as usual to see that the chop was finished, the wine drunk. Miss Barrett talked and laughed and gave no sign when anyone was in the room that she was hiding anything. Yet when they were alone she pulled out the box from under the bed and filled it hastily, stealthily, listening as she did so. And the signs of strain were unmistakable. On Sunday the church bells were ringing. "What bells are those?" somebody asked. "Marylebone Church bells", said Miss Henrietta. Miss Barrett, Flush saw, went deadly white. But nobody else seemed to notice anything.

So Monday passed, and Tuesday and Wednesday and Thursday. Over them all lay a blanket of silence, of eating and talking and lying still on the sofa as usual. Flush, tossing in uneasy sleep, dreamt that they were couched together under ferns and leaves in the darkness, in a vast forest; then the leaves were parted and he woke. It was dark; but he saw Wilson come stealthily into the room, and take the box from beneath the bed and quietly carry it outside. This was on Friday night, the 18th of September. All Saturday morning he lay as one lies who knows that at any moment now a handkerchief may drop, a low whistle may sound and the signal will be given for death or for life. He watched Miss Barrett dress herself. At a quarter to four the door opened and Wilson came in. Then the signal was given—Miss Barrett lifted him in her arms. She rose and walked to the door. For a moment they stood looking round the room. There was the sofa and by it Mr. Browning's armchair. There were the busts and the tables. The sun filtered through the ivy leaves and the blind with peasants walking blew gently out. All was as usual. All seemed to expect a million more such movements to come to them; but to Miss Barrett and Flush this was the last. Very quietly Miss Barrett shut the door.

Very quietly they slipped downstairs, past the

drawing-room, the library, the dining-room. All looked as they usually looked; smelt as they usually smelt; all were quiet as if sleeping in the hot September afternoon. On the mat in the hall Catiline lay sleeping too. They gained the front door and very quietly turned the handle. A cab was waiting outside.

"To Hodgson's", said Miss Barrett. She spoke almost in a whisper. Flush sat on her knee very still. Not for anything in the whole world would he have broken that tremendous silence.

8 *For the Love of a Man*

JACK LONDON

WHEN John Thornton froze his feet in the previous December, his partners had made him comfortable and left him to get well, going on themselves up the river to get out a raft of saw-logs for Dawson. He was still limping slightly at the time he rescued Buck, but with the continued warm weather even the slight limp left him. And here, lying by the river bank through the long spring days, watching the running water, listening lazily to the songs of birds and the hum of nature, Buck slowly won back his strength.

A rest comes very good after one has travelled three thousand miles, and it must be confessed that Buck waxed lazy as his wounds healed, his muscles swelled out, and the flesh came back to cover his bones. For that matter, they were all loafing—Buck, John Thornton, and Skeet and Nig—waiting for the raft to come that was to carry them down to Dawson. Skeet was a little Irish setter who early made friends with Buck, who, in a dying condition, was unable to resent her first advances. She had the doctor trait which some dogs possess; and as a mother cat washes her kittens, so she washed and cleansed Buck's wounds. Regularly, each morning after he had finished his breakfast, she performed her self-appointed task, till he came to look for her ministrations as much as he did for Thornton's. Nig, equally friendly, though less demonstrative, was a huge black dog, half bloodhound and half deerhound, with eyes that laughed and a boundless good nature.

To Buck's surprise these dogs manifested no jealousy towards him. They seemed to share the kindliness and largeness of John Thornton. As Buck grew stronger they enticed him into all sorts of ridiculous games, in which Thornton himself could not forbear to join; and in this fashion Buck romped through his convalescence and into a new existence. Love, genuine passionate love, was his for the first time. This he had never experienced at Judge Miller's down in the sun-kissed Santa Clara Valley. With the Judge's sons, hunting and tramping, it had been a working partnership; with the Judge's grandsons, a sort of pompous guardianship; and with the Judge himself, a stately and dignified friendship. But love that was feverish and burning, that was adoration, that was madness, it had taken John Thornton to arouse.

This man had saved his life, which was something; but, further, he was the ideal master. Other men saw to the welfare of their dogs from a sense of duty and business expediency; he saw to the welfare of his as if they were his own children, because he could not help it. And he saw further. He never forgot a kindly greeting or a cheering word, and to sit down for a long talk with them ("gas" he called it) was as much his delight as theirs. He had a way of taking Buck's head roughly between his hands, and resting his own head upon Buck's, of shaking him back and forth, the while calling him ill names that to Buck were love names. Buck knew no greater joy than that rough embrace and the sound of murmured oaths, and at each jerk back and forth it seemed that his heart would be shaken out of his body so great was its ecstasy. And when, released, he sprang to his feet, his mouth laughing, his eyes eloquent, his throat vibrant with unuttered sound, and in that fashion remained without movement, John Thornton would reverently exclaim, "God! you can all but speak!"

Buck had a trick of love expression that was akin to hurt. He would often seize Thornton's hand in his mouth and close so fiercely that the flesh bore the impress of his teeth for some time afterwards. And as Buck understood the oaths to be love words, so the man understood this feigned bite for a caress.

For the most part, however, Buck's love was expressed in adoration. While he went wild with happiness when Thornton touched him or spoke to him, he did not seek these tokens. Unlike Skeet, who was wont to shove her nose under Thornton's hand and nudge and nudge until petted, or Nig, who would stalk up and rest his great head on Thornton's knee, Buck was content to adore at a distance. He would lie by the hour, eager, alert, at Thornton's feet, looking up into his face, dwelling upon it, studying it, following with keenest interest each fleeting expression, every movement or change of feature. Or, as chance might have it, he would lie farther away, to the side or rear, watching the outlines of the man and the occasional movements of his body. And often, such was the communion in which they lived, the strength of Buck's gaze would draw John Thornton's head around, and he would return the gaze, without speech, his heart shining out of his eyes as Buck's heart shone out.

For a long time after his rescue, Buck did not like Thornton to get out of his sight. From the moment he left the tent to when he entered it again, Buck would follow at his heels. His transient masters since he had come into the Northland had bred in him a fear that no master could be permanent. He was afraid that Thornton would pass out of his sight as Perrault and François and the Scotch half-breed had passed out. Even in the night, in his dreams, he was haunted by this fear. At such times he would shake off sleep and creep through the chill to the flap of the tent, where he would stand and listen to the sound of his master's breathing.

But in spite of this great love he bore John Thornton, which seemed to bespeak the soft civilizing influence, the strain of the primitive, which the Northland had aroused in him, remained alive and active. Faithfulness and devotion, things born of fire and roof, were his; yet he retained his wildness and wiliness. He was a thing of the wild, come in from the wild to sit by John Thornton's fire, rather than a dog of the soft Southland stamped with the marks of generations of civilization. Because of his very great love, he could not steal from this man, but from any other man, in any other camp, he did not hesitate an instant; while the cunning with which he stole enabled him to escape detection.

His face and his body were scored by the teeth of many dogs, and he fought as fiercely as ever and more shrewdly. Skeet and Nig were too good-natured for quarrelling—besides, they belonged to John Thornton; but the strange dog, no matter what the breed or valour, swiftly acknowledged Buck's supremacy or found himself struggling for life with a terrible antagonist. And Buck was merciless. He had learned well the law of club and fang, and he never forwent an advantage or drew back from a foe he had started on the way to Death. He had lessoned from Spitz, and from the chief fighting dogs of the police and mail, and knew there was no middle course. He must master or be mastered;

while to show mercy was a weakness. Mercy did not exist in the primordial life. It was misunderstood for fear, and such misunderstandings made for death. Kill or be killed, eat or be eaten, was the law; and this mandate, down out of the depths of Time, he obeyed.

He was older than the days he had seen and the breaths he had drawn. He linked the past with the present, and the eternity behind him throbbed through him in a mighty rhythm to which he swayed as the tides and seasons swayed. He sat by John Thornton's fire, a broad-breasted dog, white-fanged and long-furred; but behind him were the shades of all manner of dogs, half-wolves and wild wolves, urgent and prompting, tasting the savour of the meat he ate, thirsting for the water he drank, scenting the wind with him, listening with him and telling him the sounds made by the wild life in the forest, dictating his moods, directing his actions, lying down to sleep with him when he lay down, and dreaming with him and beyond him and becoming themselves the stuff of his dreams.

So peremptorily did these shades beckon him, that each day mankind and the claims of mankind slipped farther from him. Deep in the forest a call was sounding, and as often he heard this call, mysteriously thrilling and luring, he felt compelled to turn his back upon the fire and the beaten earth around it, and to plunge into the forest, and on and on, he knew not where or why; nor did he wonder where or why, the call sounding imperiously, deep in the forest. But as often as he gained the soft unbroken earth and the green shade, the love for John Thornton drew him back to the fire again.

Thornton alone held him. The rest of mankind was as nothing. Chance travellers might praise or pet him; but he was cold under it all, and from a too demonstrative man he would get up and walk away. When Thornton's partners, Hans and Pete, arrived on the long-expected raft, Buck refused to notice them till he learned they were close to Thornton; after that he tolerated them in a passive sort of way, accepting favours from them as though he favoured them by accepting. They were of the same large type as Thornton, living close to the earth, thinking simply and seeing clearly; and ere they swung the raft into the big eddy by the saw-mill at Dawson, they understood Buck and his ways, and did not insist upon an intimacy such as obtained with Skeet and Nig.

For Thornton, however, his love seemed to grow and grow. He, alone among men, could put a pack upon Buck's back in the summer travelling. Nothing was too great for Buck to do, when Thornton commanded. One day (they had grub-staked themselves from the proceeds of the raft and left Dawson for the head-waters of the Tanana) the men and dogs were sitting on the crest of a cliff which fell away, straight down, to naked bed-rock three hundred feet below. John Thornton was sitting near the edge, Buck at his shoulder. A thoughtless whim seized Thornton, and he drew the attention of Hans and Pete to the experiment he had in mind, "Jump, Buck!" he commanded, sweeping his arm out and over the chasm. The next instant he was grappling with Buck on the extreme edge, while Hans and Pete were dragging them back to safety.

"It's uncanny," Pete said, after it was over and they had caught their speech.

Thornton shook his head. "No, it is splendid, and it is terrible, too. Do you know, it sometimes makes me afraid."

"I'm not hankering to be the man that lays hands on you while he's around," Pete announced conclusively, nodding his head towards Buck.

"Py Jingo!" was Hans's contribution. "Not mineself either."

It was at Circle City, ere the year was out, that Pete's apprehensions were realized. "Black" Burton, a man evil-tempered and malicious, had been picking a quarrel with a tenderfoot at the bar, when Thornton stepped good-naturedly between. Buck, as was his custom, was lying in a corner, head on paws, watching his master's every action. Burton struck out, without warning, straight from the shoulder. Thornton was sent spinning, and saved himself from falling only by clutching the rail of the bar.

Those who were looking on heard what was neither bark nor yelp, but a something which is best described as a roar, and they saw Buck's body rise up in the air as he left the floor for Burton's throat. The man saved his life by instinctively throwing out his arm, but was hurled backward to the floor with Buck on top of him. Buck loosed his teeth from the flesh of the arm and drove in again for the throat. This time the man succeeded only in partly blocking, and his throat was torn open. Then the crowd was upon Buck, and he was driven off; but while a surgeon checked the bleeding, he prowled up and down, growling furiously, attempting to rush in, and being forced back by an array of hostile clubs. A "miners' meeting", called on the spot, decided that the dog had sufficient provocation, and Buck was discharged. But his reputation was made, and from that day his name spread through every camp in Alaska.

Later on, in the fall of the year, he saved John Thornton's life in quite another fashion. The three partners were lining a long and narrow poling-boat down a bad stretch of rapids on the Forty-Mile Creek. Hans and Pete moved along the bank, snubbing with a thin Manila rope from tree to tree, while Thornton remained in the boat, helping its descent by means of a pole, and shouting directions to the shore. Buck, on the bank, worried and anxious, kept abreast of the boat, his eyes never off his master.

At a particularly bad spot, where a ledge of barely submerged rocks jutted out into the river, Hans cast off the rope, and, while Thornton poled the boat out into the stream, ran down the bank with the end in his hand to snub the boat when it had cleared the ledge. This it did, and was flying down-stream in a current as swift as a mill-race, when Hans checked it with the rope and checked too suddenly. The boat flirted over and snubbed in to the bank bottom up, while Thornton, flung sheer out of it, was carried down-stream towards the worst part of the rapids, a stretch of wild water in which no swimmer could live.

Buck had sprung in on the instant; and at the end of three hundred yards, amid a mad swirl of water, he overhauled Thornton. When he felt him grasp his tail, Buck headed for the bank, swimming with all his splendid strength. But the progress shoreward was slow; the progress down-stream amazingly rapid. From below came the fatal roaring where the wild current went wilder and was rent in shreds and spray by the rocks which thrust through like the teeth of an enormous comb. The suck of the water as it took the beginning of the last steep pitch was frightful, and Thornton knew that the shore was impossible. He scraped furiously over a rock, bruised across a second, and struck a third with crushing force. He clutched its slippery top with both hands, releasing Buck, and above the roar of the churning water shouted: "Go, Buck! Go!"

Buck could not hold his own, and swept on down-stream, struggling desperately, but unable to win back. When he heard Thornton's command repeated, he partly reared out of the water, throwing his head high, as though for a last look, then turned obediently towards the bank. He swam powerfully and was dragged ashore by Pete and Hans at the very point where swimming ceased to be possible and destruction began.

They knew that the time a man could cling to a slippery rock in the face of that driving current was a matter of minutes, and they ran as fast as they could up the bank to a point far above where Thornton was hanging on. They attached the line with which they had been snubbing the boat to Buck's neck and shoulders, being careful that it should neither strangle him nor impede his swimming, and launched him into the stream. He struck out boldly, but not straight enough into the stream. He discovered the mistake too late, when Thornton was abreast of him and a bare half-dozen strokes away while he was being carried helplessly past.

Hans promptly snubbed with the rope, as though Buck were a boat. The rope thus tightening on him in the sweep of the current, he was jerked under the surface, and under the surface he remained till his body struck against the bank and he was hauled out. He was half drowned, and Hans and Pete threw themselves upon him, pounding the breath into him and the water out of him. He staggered to his feet and fell down. The faint sound of Thornton's voice came to them, and though they could not make out the words of it, they knew that he was in his extremity. His master's voice acted on Buck like an electric shock. He sprang to his feet and ran up the bank ahead of the men to the point of his previous departure.

Again the rope was attached and he was launched, and again he struck out, but this time straight into the stream. He had miscalculated once, but he would not be guilty of it a second time. Hans paid out the rope, permitting no slack, while Pete kept it clear of coils. Buck held on till he was on a line straight above Thornton; then he turned, and with the speed of an express train headed down upon him. Thornton saw him coming, and, as Buck struck him like a battering ram, with the whole force of the current behind him, he reached up and closed with both arms around the shaggy neck. Hans snubbed the rope around the tree, and Buck and Thornton were jerked under the water. Strangling, suffocating, sometimes one uppermost and sometimes the other, dragging over the jagged bottom, smashing against rocks and snags, they veered in to the bank.

Thornton came to, belly downward and being violently propelled back and forth across a drift log by Hans and Pete. His first glance was for Buck, over whose limp and apparently lifeless body Nig was setting up a howl, while Skeet was licking the wet face and closed eyes. Thornton was himself bruised and battered, and he went carefully over Buck's body, when he had been brought around, finding three broken ribs.

"That settles it," he announced. "We camp right here." And camp they did, till Buck's ribs knitted and he was able to travel.

That winter, at Dawson, Buck performed another exploit, not so heroic, perhaps, but one that put his name many notches higher on the totem-pole of Alaskan fame. This exploit was particularly gratifying to the three men; for they stood in need of the outfit which it furnished, and were enabled to make a long-desired trip into the virgin East, where miners had not yet appeared. It was brought about by a conversation in the Eldorado Saloon, in which men waxed boastful of their favourite dogs. Buck, because of his record, was the target for these men, and Thornton was driven stoutly to defend him. At the end of half an hour one man stated that his dog could start a sled with five hunded pounds and walk off with it; a second bragged six hundred for his dog; and a third, seven hundred.

"Pooh! Pooh!" said John Thornton; "Buck can start a thousand pounds."

"And break it out? And walk off with it for a hundred yards?" demanded Matthewson, a Bonanza King, he of the seven hundred vaunt.

"And break it out, and walk off with it for a hundred yards," John Thornton said coolly.

"Well," Matthewson said, slowly and deliberately, so that all could hear, "I've got a thousand dollars that says he can't. And there it is." So saying, he slammed a sack of gold dust of the size of a bologna sausage down upon the bar.

Nobody spoke. Thornton's bluff, if bluff it was, had been called. He could feel a flush of warm blood creeping up his face. His tongue had tricked him. He did not know whether Buck could start a thousand pounds. Half a ton! The enormousness of it appalled him. He had great faith in Buck's strength and had often thought him capable of starting such a load; but never, as now, had he faced the possibility of it, the eyes of a dozen men fixed upon him, silent and waiting. Further, he had no thousand dollars; nor had Hans or Pete.

"I've got a sled standing outside now, with twenty fifty-pound sacks of flour on it," Matthewson went on with brutal directness; "so don't let that hinder you."

Thornton did not reply. He did not know what to say. He glanced from face to face in the absent way of a man who has lost the power of thought and is seeking somewhere to find the thing that will start it going again. The face of Jim O'Brien, a Mastodon King and old-time comrade, caught his eyes. It was as a cue to him, seeming to rouse him to do what he would never have dreamed of doing.

"Can you lend me a thousand?" he asked, almost in a whisper.

"Sure," answered O'Brien, thumping down a plethoric sack by the side of Matthewson's. "Though it's little faith I'm having, John, that the beast can do the trick."

The Eldorado emptied its occupants into the street to see the test. The tables were deserted, and the dealers and gamekeepers came forth to see the outcome of the wager and to lay odds. Several hundred men, furred and mittened, banked around the sled within easy distance. Matthewson's sled, loaded with a thousand pounds of flour, had been standing for a couple of hours, and in the intense cold (it was sixty below zero) the runners had frozen fast to the hard-packed snow. Men offered odds of two to one that Buck could not budge the sled. A quibble arose concerning the phrase "break out". O'Brien contended it was Thornton's privilege to knock the runners loose, leaving Buck to "break it out" from a dead standstill. Matthewson insisted that the phrase included breaking the runners from the frozen grip of the snow. A majority of the men who had witnessed the making of the bet decided in his favour, whereat the odds went up to three to one against Buck.

There were no takers. Not a man believed him capable of the feat. Thornton had been hurried into the wager, heavy with doubt; and now that he looked at the sled itself, the concrete fact, with the regular team of ten dogs curled up in the snow before it, the more impossible the task appeared. Matthewson waxed jubilant.

"Three to one!" he proclaimed. "I'll lay you another thousand at that figure, Thornton. What d'ye say?"

Thornton's doubt was strong in his face, but his fighting spirit was aroused—the fighting spirit that soars above odds, fails to recognize the impossible, and is deaf to all save the clamour for battle. He called Hans and Pete to him. Their sacks were slim, and with his own the three partners could rake together only two hundred dollars. In the ebb of their fortunes, this sum was their total capital; yet they laid it unhesitatingly against Matthewson's six hundred.

The team of ten dogs was unhitched, and Buck, with his own harness, was put in the sled. He had caught the contagion of the excitement, and he felt that in some way he must do a great thing for John Thornton. Murmurs of admiration at his splendid appearance went up. He was in perfect condition, without an ounce of superfluous flesh, and the one hundred and fifty pounds that he weighed were so many pounds of grit and virility. His furry coat shone with the sheen of silk. Down the neck and across the shoulders, his mane, in repose as it was, half bristled and seemed to lift with every movement, as though excess of vigour made each particular hair alive and active. The great breast and heavy forelegs were no more than in proportion with the rest of the body, where the muscles showed in tight rolls underneath the skin. Men felt these muscles and proclaimed them hard as iron, and the odds went down to two to one.

"Gad, sir! Gad sir!" stuttered a member of the latest dynasty, a king of the Skookum Benches. "I offer you eight hundred for him, sir, before the test, sir; eight hundred just as he stands."

Thornton shook his head and stepped to Buck's side.

"You must stand off from him," Matthewson protested. "Free play and plenty of room."

The crowd fell silent; only could be heard the voices of the gamblers vainly offering two to one. Everybody acknowledged Buck a magnificent animal, but twenty fifty-pound sacks of flour bulked too large in their eyes for them to loosen their pouch-strings.

Thornton knelt down by Buck's side. He took his head in his two hands and rested cheek on cheek. He did not playfully shake him, as was his wont, or murmur soft love curses; but he whispered in his ear. "As you love me, Buck. As you love me," was what he whispered. Buck whined with suppressed eagerness.

The crowd was watching curiously. The affair was growing mysterious. It seemed like a conjuration. As Thornton got to his feet, Buck seized his mittened hand between his jaws, pressing in with his teeth and releasing slowly, half-reluctantly. It was the answer, in terms, not of speech, but of love. Thornton stepped well back.

"Now, Buck," he said.

Buck tightened the traces, then slackened them for a matter of several inches. It was the way he had learned.

"Gee!" Thornton's voice rang out, sharp in the tense silence.

Buck swung to the right ending the movement in a plunge that took up the slack and with a sudden jerk arrested his one hundred and fifty pound. The load quivered, and from under the runners arose a crisp crackling.

"Haw!" Thornton commanded.

Buck duplicated the manoeuvre, this time to the left. The crackling turned into a snapping, the sled pivoting and the runners slipping and grating several inches to the side. The sled was broken out. Men were holding their breaths, intensely unconscious of the fact.

"Now, MUSH!"

Thornton's command cracked out like a pistol-shot. Buck threw himself forward, tightening the traces with a jarring lunge. His whole body was gathered compactly together in the tremendous effort, the muscles writhing and knotting like live things under the silky fur. His great chest was low to the ground, his head forward and down, while his feet were flying like mad, the claws scarring the hard-packed snow in parallel grooves. The sled swayed and trembled, half-started forward. One of his feet slipped, and one man groaned aloud. Then the sled lurched ahead in what appeared a rapid succession of jerks, though it never really came to a dead stop again . . . half an inch . . . an inch . . . two inches. . . .

The jerks perceptibly diminished; as the sled gained momentum, he caught them up, till it was moving steadily along.

Men gasped and began to breath again, unaware that for a moment they had ceased to breathe. Thornton was running behind, encouraging Buck with short, cheery words. The distance had been measured off, and as he neared the pile of firewood which marked the end of the hundred yards, a cheer began to grow and grow, which burst into a roar as he passed the firewood and halted at command. Every man was tearing himself loose, even Matthewson. Hats and mittens were flying in the air. Men were shaking hands, it did not matter with whom, and bubbling over in a general incoherent babel.

But Thornton fell on his knees beside Buck. Head was against head, and he was shaking him back and forth. Those who hurried up heard him cursing Buck, and he cursed him long and fervently, and softly and lovingly.

"Gad, sir! Gad, sir!" spluttered the Skookum Bench king. "I'll give you a thousand for him, sir, a thousand, sir—twelve hundred, sir!"

Thornton rose to his feet. His eyes were wet. The tears were streaming frankly down his cheeks. "Sir," he said to the Skookum Bench king, "no, sir. You can go to hell, sir. It's the best I can do for you, sir."

Buck seized Thornton's hand in his teeth. Thornton shook him back and forth. As though animated by a common impulse, the onlookers drew back to a respectful distance; nor were they again indiscreet enough to interrupt.

9 *Training a Police Dog*

ARTHUR HOLMAN

WE always had a dog at home—a lurcher, a terrier, or a collie—so it could be said that I grew up with them. After leaving school at fourteen my first jobs were with dogs, going out with the sheep- and cattle-dogs owned by local farmers.

Oddly, Rex III was not only the third police dog of that name, but the third dog I had owned called Rex. The first was a collie, which I bought from a gipsy for five shillings when I was seventeen. I do not know where the gipsy obtained Rex or how she had treated him, but it is literally true that Rex could smell gipsies half a mile off. When he did there was no holding him. Barking furiously, he would race down to the front gate and growl and snarl and bare his teeth until they had passed and were out of sight.

On May 10, 1932, aged twenty-one, I joined the Metropolitan Police. I was not motivated by any romantic notions of becoming a world-acclaimed super sleuth, but by the Depression. It was the Police or the Army. Rex, alas, had to be left behind with my father, since dogs were not permitted in section houses—police bachelor quarters.

I bought Rex II, a fine black Labrador, soon after my marriage to the girl with whom I had gone to school, and he stayed with us until the outbreak of war. When the air-raids started my family was evacuated, and the people they went to had no accommodation for a pet. I thought it would be unfair to leave Rex alone in a blitz-threatened house all day while I was on duty, and decided, reluctantly, to give him to relatives in the country.

After the War I was attached to the C.I.D. as an *aide* to the Ghost Squad, carrying out undercover observations in the Tooting, Balham, and Mitcham areas. Then, early in 1950, Scotland Yard asked for volunteers who had previous experience with dogs. A fairly new, exciting experiment was to be extended—the training of dogs to assist in the detection of crime. There was no need for me to consider the request; I immediately put down my name.

I was interviewed by a chief inspector at Tooting police station (to which I was attached), and he forwarded my name and qualifications to Whitehall. Soon afterwards, on April 12, the chief inspector told me I could collect my dog next day from Imber Court, near Esher. Collar and lead? He knew nothing about them. I dug into the cupboard for those belonging to Rex II.

On the following morning a police van took me the eight miles from Mitcham to Esher. There I met two other policemen who had come to collect dogs—Sergeant Baynton and P.C. MacCallum—and the three of us were taken by Sergeant Boath to the kennels.

"The three dogs," said Boath, "are Kim, Roy, and Rex Three."

"Rex Three?" I said. "Well, I'm darned! That's the one for me."

"He's the oldest of them," warned Boath.

The coincidence was remarkable, and I was determined, even before seeing him, that Rex III would be mine. But what of the others? As we walked along I told them how foolish I had been to say I wanted Rex. The older the dog the harder he was to train. . . . Beyond the puppy stage untrained dogs were intractable and savage. . . . Whoever chose Rex would be taking on a dangerous handful—and so one. My ruse worked. When we arrived at the kennels Baynton and MacCallum made for Kim and Roy. Rex III was mine.

As soon as I saw him I knew I had not made a mistake. To people who have never owned an animal a dog is simply a dog; there is no adjectival qualification or distinction. All the descriptive poetry in the world could not convey to them how I felt at that moment—the thrill of tingling excitement that went through me when I saw that Rex was indeed a king among dogs. An Alsatian, big for his twelve months, with black-and-tan markings, clear brown eyes, and sharply defined, intelligent features, he had a vital force I have never known in another dog, before or since.

I stroked him for a while, spoke to him so that he would recognize my voice, then put on his collar and lead and took him for our first walk together. Happy to be away from his kennel, Rex bounded through the grass, and such was his strength he dragged me a few yards before I was able to slow him to walking-pace. Even then he tugged at the lead, and I spent the rest of that day trying to teach him to walk to "Heel"—to walk with his head close to my left knee. But as soon as I forced him back, so he strained forward again. By three o'clock that afternoon my hands were raw and my arms aching—and still Rex would not walk to "Heel."

Returning to the kennels, I met Tom Scott, a police constable from Staines who was at Imber training his own dog, Chum. "This dog," I said, "is as hard and fit as they come. He's all bone and muscle—and look at the condition of his coat. It's perfect."

Scott examined Rex with an expert eye. "That dog is going to be a cracker!"

"I think you're right," I said. "It's my thirty-ninth birthday today, and I couldn't have had a better present. . . . If he *is* mine!"

Before leaving I asked Sergeant Boath about the dog's licence and ownership. "Catch any thieves with him," he assured me, "and he's yours." (We did catch thieves —Rex was to make a total of a hundred and twenty-five arrests—and although his licence was held by the Metropolitan Police, he was my dog, for he lived with me, was trained and handled by me, operated only with me, would never eat unless fed by me, and would not obey a single order if I did not give it.) I was told, too, that although having Rex meant no extra pay, the police would provide him with a pound and a half of meat and a pound of biscuits a day, and a kennel. It was hoped that the latter had been sent to Mitcham and would be awaiting me at my home.

Rex, I learned, had been born on March 11, 1949, and had a formidable pedigree. As a puppy he had kept the watchman at the French Legation company at night, but he fell a victim to automation. The Legation installed automatic burglar alarms, and in March 1950 Rex had been given to the Metropolitan Police.

As we climbed into the van Rex dived under the seat. I wondered if he was nervous of travelling, but when I looked down I could see that he was quite happy. We stopped at Putney police station for his meat, and were home by five o'clock.

The family knew I was bringing home a dog, and a reception committee was waiting when I opened the front door—my wife, Betty; my daughter, Vivienne, then aged ten; and my son, Richard, fourteen. When they saw Rex their expressions froze.

Betty: "I expected a Labrador!"

Vivienne: "I wanted a new-born puppy!"

Richard: "I don't think much of that!"

I had one answer for the three of them: "Police orders."

Knowing that I would have to start Rex on a regular routine, I fed him, took him for a run, then tied him up for the night. He settled down right away, and there were no barks or whimpers to disturb my sleep. Next morning he seemed pleased to see me. We went into the garden, and I secured his lead to a fence before preparing the family's tea. After breakfast I groomed him with the brush and comb I had kept since the days of Rex II.

After those first three nights three years were to pass before Rex was allowed in the house again, even for his meals. There was nothing cruel about this, for he had a natural coat to protect him against the weather, and it would have been less humane to have kept him *in* the house on winter nights if we were called out on a job I could adjust myself from the warmth to the outside temperature by a scarf, gloves, hat, and overcoat; Rex could not. He moved indoors, finally, because I was afraid of complaints from neighbours. When colleagues called for us at night I was awakened by the dog's barking long before the knock came on the door, and was certain that, although they did not say anything, the neighbours were aroused too. I did not take him indoors earlier because, despite my wife's fondness for Rex, it took her all that time to lose her mistrust of Alsatians. (Rex was not bothered by non-human sounds in the night, and never barked during the day at *regular* callers. I emphasize the word "regular," for despite his indifference towards tradesmen, if one called on a day other than his usual one Rex would bark furiously. I never fathomed his curious ability for distinguishing not only between people, but between days of the week; yet he did have a sort of "calendar memory.")

Every day a van was to take us to Imber Court and bring us home again, and my wife found some compensation for having Rex about the house in the knowledge that for three months I would keep "office hours" and miss night duty.

I spent the whole of our second day at Imber in

further attempts to teach Rex to walk to "Heel." My arms ached dreadfully, and during the afternoon he pulled so hard as I tried to drag him back that the lead broke. I borrowed another from Tom Scott, but it was a waste of time. I grew despondent, and felt like throwing in my hand. If he won't obey a simple rule like this, I thought, I'll never be able to train him.

That evening I bought a small chain choke collar. Despite its name, it is in no way cruel—its use is approved by the Kennel Club—and works on the principle that when the collar becomes too tight the dog will stop pulling. But Rex seemed to be made of iron, and tugged at the lead as powerfully as before. (He was never to walk to "Heel" on a lead. Ironically, however, he did so when not attached to it, and as I seldom had him on the lead once his training had ended, this solitary defect became a purely academic one.)

The next step was to get Rex to sit by my side when I stopped walking. I taught him to do this by holding the lead in my right hand and pushing his rear down with my left. After only an hour's practice he understood perfectly, and whenever I halted, with or without the lead, he would automatically come to the "Sit" position.

I experimented with him "doing a chase", Scott, wearing protective clothing, played the supposed criminal, and at his first attempt Rex knew what was expected of him. The words "Stop him!" and a wave of my hand sent Rex bounding after Scott. He caught the policeman's arm, and despite my colleague's efforts to shake free, Rex would not let go until I went up and pulled him away, saying "Leave" as I did so. Repetition taught Rex the meaning of this command. "My word," said Scott, "he does come in! He comes in fast, and he holds hard. No one will get away from you while you've got Rex with you."

No one ever did.

Although my dog was an expert chaser, it would never do if the order "Stop him!" sent Rex after the wrong person. It would prove both embarrassing and dangerous if he attacked a policeman pursuing a suspect, pounced upon a man running for a bus, or leapt on a child rushing home from school.

I found, however, that Rex had an uncanny gift for quickly recognising friends in and out of uniform. When we were on a job in company with other policemen, and one of them started running after a criminal, the order "Stop him!" sent Rex straight past my colleague in chase of the wanted man. In any case, on the occasions when we were out alone there was seldom more than one running person in sight when the order was given, and if Rex *did* make a mistake the command "No!" was enough for him to leave that person alone and continue after the correct one.

I recall a night when Rex and I were in a police car patrolling Battersea looking for a man who had taken part in a smash-and-grab raid. I glimpsed the man running towards a side-turning—he was carrying twelve raincoats, which I thought an unduly large number for a cloudless night—and the order "Stop him!" sent Rex through the open window of the car before I could open the door. By a freak of coincidence another man came running out of the night diagonally across the path of the man with the macs, and Rex, put off by the appearance of this second person, jumped at him. But my roared "No!" made him swerve away after the real criminal, whom he caught.

My work would have been wasted if Rex obeyed other people's commands. If, for instance, I told Rex to stop a criminal, yet he took notice when the prisoner, knowing the jargon, said, "Leave . . . Sit," we would be faced with a situation bordering on the farcical. So I trained him to obey only my voice by administering a verbal rebuke when he took heed of anyone else's. To help avoid confusion if there was more than one handler and dog around, I also changed the words of a number of commands. Thus I would say "Hand" while other handlers employed "Come," and Rex, not understanding the latter word, would take no notice, even although he might have muddled our voices.

Making Rex respond to "Hand" was a severe test on my patience—and the impatient person will never fully train a dog. During this part of his training, on April 26, a day when there was a heavy, freak snowstorm, Rex ran away from me, despite my shouts of "Hand." I stumbled after him, soaked through and half-frozen. It was perhaps an hour before he trotted back to me.

Next day when I let him off the lead he again ran off. I tried to coax him back with pieces of meat I had put in my pocket for this purpose. He took no notice.

When it happened on the third successive day I asked Tom Scott to lend me his catapult and a couple of acorns. As a boy I had been a good shot with a catapult, and I discovered that my skill was still with me. I scored a hit first try, and as the acorn struck Rex's rear I shouted "Hand!" He was so surprised to discover I could reach him at such a distance that he stopped dead —then hurried to my side. From that moment I had no further trouble. Whatever he was doing, the command "Hand" brought him quickly to me.

A point for dog-owners:

No dog worthy of his biscuits likes to be on a lead, and I always put the lead out of sight when calling Rex. I did not want him to think he was going to lose his freedom by obeying my order. I know there is nothing more infuriating than to call a dog and not to have him come, and although your instinct might be to beat the dog for his disobedience, I believe strongly that this is a monstrous injustice to the animal. Dogs cannot reason as we do, and from his point of view he is being punished for coming to you—for obeying your order. When the dog ultimately comes he should be fussed over. He will grow to believe that you are calling him to pet him, and the chances are that soon he will come quickly whenever you tell him to. I feel, too, that it is cruel to the dog to have him tugging at the end of a lead which is pulling against his windpipe. Far better, surely, to train a dog to walk to "Heel".

Since there were bound to be times when silence was vital, and I would be unable to call him, another trick of mine ensured that when we were tracking and seeking Rex did not run too far ahead. I would duck behind a tree and lie flat on the ground. When Rex looked round he found I had disappeared and came to search for me. In time he never ran in front of me without continually glancing backward to make sure I was still there, and he never kept more than thirty yards ahead of me, unless specifically ordered away. Scores of times people were to remark, "How that dog watches you! He must really love you. He keeps turning round to see no one touches you. How did you train him to get like that?" Now they know.

Another variation of words I employed was "Find him," whereas other handlers said "Seek". This order as it implies, meant letting Rex loose to find a hidden suspect. At first he was not too good at it. Time—and patience—were required before he grasped that "Find him" contained three commands in one—to discover the man we were after, to guide me to him by barking non-stop, and to prevent the suspect from escaping.

At the beginning I would say, "Find him," and take Rex on the lead to where the person was hiding. I repeated this over and over again, until Rex grasped what was wanted and went sniffing and searching without the lead. There was no risk of Rex's finding the wrong person (someone having a quiet nap in the shade of a tree, or a courting couple), because he would be used for seeking only at places where no one had any right to be—in a warehouse at night, for instance, or in the fenced grounds of a factory.

Once he had located the hidden policeman acting as a suspect, I would tell Rex to "Speak." (This is something anyone can teach the average dog to do. Hold a piece of food just beyond his reach and keep repeating "Speak." When the teasing makes the dog bark, give him the food. Next time say "Speak" first, then start the teasing and the repetition of the word until he again barks. In time he will understand he must bark for the food, and that barking is associated with the word "Speak." Finally, the word "Speak" will produce a bark—in the hope of being given food.) If Rex stopped barking the "suspect" teased him until he resumed, and constant practice brought us to the stage when "Find him!" sent Rex bounding off in search of a suspect and, having found him, barking until I came along and told him to be quiet.

The last part of the lesson was for Rex not to allow the suspect to run away, and I taught him to do this by telling my colleague to make a run for it and commanding Rex, "Stop him!" Ultimately he had it perfect: that implicit in "Find him!" was to discover the wanted man, bark until I arrived, and prevent the suspect from escaping. Naturally, it was not as straightforward as it reads here. A dog, even a highly intelligent one, is still governed by basic limitations, and there were times when it was all I could do to contain my impatience. But I knew that if I lost my temper with Rex I would destroy all I had achieved, including the most vital element—his faith in me. As it was, Rex became probably the best seeker in the country.

Since my colleagues acted as suspects while I was training Rex, I did the same for them. One day I played the criminal when Scott was teaching Chum to "Find". The dog discovered my hiding-place—and promptly bit me in the neck. Fortunately his teeth barely grazed the skin.

On another occasion the policeman after whom Rex was sent hid in a tree. Regrettably he did not climb high enough. The dog leapt at him, and when I came along he was swinging like a pendulum, his teeth hard into the seat of his victim's trousers. I doubled up with laughter, but my colleague did not share the joke.

Rex was no mean jumper and hurdler. At first I took him on his lead to a small obstacle and helped him over with the word "Up". We went on to a higher obstruction, and then to one even higher. In time the command "Up" sent him over barriers unaided, and eventually I could shout "Up" and Rex would cross a dozen obstacles spaced irregularly over a distance of a hundred yards. He took his eighty-five pounds effortlessly over a fence or a wall eight feet high.

Many people think that training dogs is a cruel, unnatural business. They say, "Dogs were not meant to behave in this manner, and it is wicked to force them to do things against their true inclination." Such arguments are fallacious. They may apply to circus acts, where dogs are put into fancy dress, made to walk on two of their four legs, and compelled to ride miniature bicycles, but the training of police dogs is a completely different matter. For we do no more than bring out and develop a dog's intrinsic, latent talents, and teach them a discipline well within their powers of natural obedience—an ideal life for an Alsatian. Remember that we employ only those breeds of dog physically and mentally capable of the tasks set them. When talking of cruelty one must bear in mind that no one can rely upon a maltreated bullied animal disciplined by the strap and starvation, and I have yet to hear of a well-trained police dog turning against its handler, or failing to do its duty to the best of its ability.

Training was not unduly arduous. After a period of, say, "general obedience" we would break off for a run, a rest, or a drink before starting our next lesson. On the other hand, there were times when Rex needed extra cramming after school hours.

Homework was necessary, for instance, during the period I was teaching Rex to "Stay" when I walked away from him. Roy, Chum, and Kim were far better at this than Rex, and the fact peeved me, for I knew I had the best all-rounder, and I wanted Rex to be top of the form in every subject, especially since MacCallum had jokingly assured me, "I'll catch more thieves in a month than you will in a year"—words I was determined to make him eat.

The method of training was simple: I would push down his behind, tell him "Stay," then walk away. Time and again he trotted after me, and at one stage I really thought that when his passing-out came along he would fail this test of obedience. Finally I realized that Rex was being distracted by people and other dogs moving about at Imber, and I continued this part of his training in the seclusion of my back garden at home. There he quickly understood what I wanted of him, and I could leave him for a full hour without his showing any sign of restlessness. When we continued at Imber Court I pushed him down on the ground and made him remain still while people stepped around and over him. In this way he came to obey the order despite distractions.

When Rex became thirsty I took him to the drinking bowls, saying "Drink" as I did so. It was not long before he understood the word, and when I uttered it he would go racing off to slake his thirst. This item of training saved me long walks when the bowls were perhaps a quarter of a mile from our "classroom". Having had his drink, Rex always hurried back to my side.

It was important that Rex should not accept food from other people. For one thing he could be poisoned, for another he could be kicked or bludgeoned while busy eating it. So I attached him to a rope, one of my colleagues offered him food, and as Rex went to take it he was given, not the food, but a light flick at his nose. (The rope was to prevent his attacking the person who annoyed him.) As the only person who did not tease him was myself, he soon refused to have anything to do with anyone else, with or without a flick, who offered him titbits. This was true even at home.

On one occasion, when I was not present, a policeman started tormenting Rex with the food-and-flick trick, and thought it a tremendous joke. By the time I returned he had tired of his malicious teasing, and was some distance away. I released Rex for a run, and he streaked towards the other policeman. When I caught up with Rex his teeth were hard into the seat of his victim's trousers, and my colleague was screaming with both pain and terror. I ordered Rex to "Leave," and saw blood streaming from where the man had been mauled.

I was unable to fathom why Rex had attacked him, but when other policemen told me the story I felt Rex deserved a pat rather than punishment. No animal likes to be humiliated, and Rex was able to retaliate with something more than an irate lap-dog's yelp. An Alsatian's teeth make a formidable weapon.

Since Rex was to be occupied with the serious business of preventing and detecting crime, I did not like the idea of his being cajoled by clicking fingers, seductive sounds, or captivating phrases. I could not anticipate all the sounds or words, so I settled for the two most popular—"Good dog" and the kissing noise so many people employ when beckoning an unknown animal. I, too, made the sound, but when Rex came to me he was rewarded with the word "No". The same applied to "Good dog" and clicking fingers. Although Rex would not attack me, he quickly set about others who tried to entice him in this way. Since he was always close to me unless going after, or looking for, a suspect, if some unknowing passer-by tried to attract his attention by noises or gestures I would quickly say "No" before the dog went for him.

I wanted Rex to protect me if I was attacked, and I trained him to do this in a very simple manner. Although Rex was disciplined to obey orders, natural instinct made him defend himself when set upon, without any prompting from me. By asking a colleague to pretend to attack Rex when the dog was by my side, and then to threaten me while Rex was still there, I got Rex to regard an assault upon me as grimly as he did one upon himself. This held good even when he was some distance from me: a raised fist brought him barking and snarling to my defence.

It would have been awkward if Rex had attacked people who came up to me for peaceful reasons—if the superintendent felt like a chat, a pedestrian wanted to know the time, or my wife gave me a kiss. In those circumstances I told Rex to "Sit", and since he had the habit of never taking his eyes from me when sitting, I knew I had protection if the pedestrian asking the way to the High Street turned out to be an armed idiot. Passers-by who gave him a friendly pat were quite safe: he never attacked anyone unless he or I were threatened, or I ordered him to. But I could never skylark with my children when Rex was about: he would suppose I was being attacked and rush to my defence, despite orders to "Sit".

Even when he knew all these things Rex's training was far from complete. Apart from finding concealed criminals, he would need to follow the trail of suspects who had left the scene of a crime before our arrival there. This is known as tracking, and one morning, in company with Tom Scott and Chum, I took Rex to Esher Common to start teaching him how to do it. Tom tied Chum to a tree and walked out of sight through the bushes stopping about a hundred and fifty yards away. I pointed to the ground and said, "Track". Not unreasonably, Rex just stood still and looked at me inquiringly. Keeping ahead of the dog, with my hand low to the ground, and pointing with my fingers, I followed Tom's trail, saying, "Track . . . Track. . . ." My pointing fingers kept Rex's nose and attention to the ground, and he finally realized that we were following one person's scent. "Track . . . Track . . . ," I repeated, until we came to Tom crouching in the bracken.

Later that day, on Mitcham Common, I tried Rex at tracking my wife, but he was not very successful. As it transpired, all that was needed was practice; there was nothing wrong with his nose for a scent. This was proved quite early in his training, when we were tracking MacCallum at Esher and a number of people crossed the trail ahead of us. Rex, still a novice, none the less kept to the policeman's scent despite the confusion caused by the other humans. When he was fully trained Rex proved as good as any bloodhound. The most difficult tracking (and this holds for all dogs) was along concrete roads, but eventually Rex took them in his stride. Daily, the people he was sent to track were given longer and longer starts, across more and more varied terrain. Finally Rex could lead me to a person who had started off an *hour and a half* earlier.

Retrieving was also an integral part of Rex's training,

and his initial exercises at it were straightforward, but lengthy. At first I threw his lead away and said, "Fetch". He was baffled, and showed it. So I went with him to the lead, placed it in his mouth, walked away, and said again, "Fetch", beckoning him to come to me. I did this over and over again until he understood what I wanted, and "Fetch" sent him bounding away into the grass. I varied the objects after which I sent him, and I was able, ultimately, to throw a penny far into the undergrowth and have it brought back to me. I would have tried a sixpence but for the danger of his swallowing it.

This limited retrieving is enough if you are playing ball in the park, but a police dog must be trained to pick up and bring back *everything* it finds that retains a human scent, even an unfamiliar one, for a seemingly trivial object dropped by a criminal can be an important —even vital—clue.

So my next step was to say "Fetch" although I had not thrown away or hidden anything in the grass. Rex rushed off, searched around, and came back to me empty-mouthed and dejected. "Fetch," I insisted, and after only the briefest hesitant pause he hurried away again. This went on until, in desperation, he returned with a cigarette packet "planted" by one of my colleagues. I patted him to assure him he had done the correct thing, and sent him off yet again. This time he returned with a newspaper. Before long he realized that "Fetch" meant he must bring back *anything* he could find. The exceptions were items, such as stones and twigs, that clearly had not been handled by a human being, and when he came to me with them in his mouth I would say, "No"—a word even the stupidest dog can understand early on in its training. He grew to understand that I wanted only those things that still preserved a human smell.

It is a sign of weakness to threaten a dog with one's hand, a stick, or a lead. Tone of voice when saying "No" is quite sufficient to make him realize he had done something, or attempted to do something, wrong. Similarly, a kindly "Good dog" and a pat when he has done the right thing is far more practical and winning than a placatory knob of sugar prompted by remorse for a too-free slap. Remember, before you beat your animal, that a dog is probably the only friend you have who is prepared to die for you.

Spring warmed into summer, and Rex's training was well up to schedule. He could now obey the orders "Sit . . . Stop him . . . Leave . . . Hand . . . Heel . . . Speak . . . Up . . . Find . . . Stay . . . Track . . . Fetch"—and although this was a staggeringly impressive list, there were still omissions. One of these I rectified on Epsom's famous racecourse, where I taught Rex to crawl. I pushed him down on his stomach, walked a few yards ahead, then turned and said, "Crawl," beckoning Rex to me. He promptly rose to his feet. I went to him, pushed him down, and repeated the performance. The crawl was difficult for him to understand at first, but, as with everything, Rex mastered it in time. Crawling proved a most useful accomplishment. On one occasion, for example, I was to send Rex after a suspect loitering in the grounds of a factory. The chain-link fence was too high for Rex to leap over, and was topped by barbed wire, and the entrance-gate was some distance away. At the order "Crawl" Rex flattened down, wriggled under the fence—and caught the man.

I wanted Rex to grow accustomed to the sound of revolver-fire, and I arranged with Inspector Blackmore at the Imber Court small-arms range to go there once a week with Rex and walk around for half an hour when firing was in progress. The dog soon grew used to the noise. Rex was never worried by the other sort of fire (the burning variety), and a combination of the two—the bangs and blazes of Guy Fawkes' night—in no way disturbed him.

At first Rex was a reluctant swimmer. No amount of coaxing could get him into the water. I overcame this by taking him to the bank of the river Ember, ordering him to "Stay," then crossing the river by boat and calling him to hand. He went into a dreadful panic. He approached the water, backed away, barked, and ran up and down the bank. Eventually his desire to be with me was greater than his fear of the water, and he plunged in. Once initiated, he became a keen swimmer—and a champion diver. I took him to the local pool, walked with him to the edge of the diving-board, threw down a pebble, and said, "Fetch." He went in without any trouble, retrieved the stone, and brought it to me. I doubt whether many dogs could do that.

I also taught him to chase across water. The "suspect" would start running from close to the river on the opposite bank, and "Stop him!" sent Rex swimming strongly across the Ember and bounding in pursuit I would follow (via a footbridge!) and tell Rex to "Leave".

In a bathing-hut I taught Rex how to guard a prisoner. Another policeman played the part of the criminal, and I told Rex to sit in front of him. (I chose a bathing-hut because of its confined space. Rex had to sit close to my colleague, and when fully trained he always did this as a natural part of being on guard. It had a practical value: the bared, menacing teeth of an Alsatian a few inches from a criminal's throat were almost invariably sufficiently intimidating to prevent his attempting an escape.) As Rex sat there I said, "Guard," and my colleague started to tease the dog. I made sure Rex did not leap at him, but the teasing made Rex show his teeth and growl with rage. If Rex stopped growling renewed teasing restarted it. In case a prisoner did try to escape, I told my colleague to attempt to leave the bathing-hut, and ordered Rex to stop him. Eventually Rex was able to grasp the composite instructions contained in the order "Guard him," and to translate the words into sitting close to the prisoner, growling savagely, ready to pounce if he attempted to get away.

On one occasion I left Rex on guard at the bathing-hut, went off for a few minutes (as though to summon assistance), and returned to find Tom Scott in a near state of collapse. Rex had stopped growling, and the policeman made a gesture to tease him. As soon as he moved Rex pounced. His teeth ripped out the sleeves or both Tom's coat and shirt, and bit into the flesh of his shoulder. There was a dickens of a row, for my colleague was not wearing uniform, and his natty sports coat was quite beyond repair.

I used the Sports Club buildings at Imber to teach Rex how to cope with a criminal who fled into a house, slammed the door behind him, and tried to make his escape through a door or window at the other side of the premises. When the policeman being chased ran into a building Rex began to bark with frustration and rage. The policeman went to a side-window and attracted the dog's attention. As Rex ran round, my colleague ducked out of sight and reappeared somewhere else. This continued until Rex had circled the building a number of times, and the point was reached when Rex automatically chased round and round a house in which the suspect had taken refuge, turning the building into a self-made prison.

There was a lighter side to Rex's training. I would confound people who were watching him eat by saying, "I wouldn't eat that; it's bad," whereupon he immediately dropped the food from his mouth. I had taught him to do this by holding his head, taking from his mouth whatever he was eating, and saying, "Bad". After my practising with him only half a dozen times he had it off pat, and the word "Bad" made him expectorate what was in his mouth.

The only other purely party piece I taught him was the simple one of shaking hands. Again, it involved no more than a certain amount of practice—to lift his paw and shake it, at the same time repeating, "Say how do you do," until the words caused him to raise his paw whenever he heard me say them.

By this time Rex was fully trained, but there was one extra-curricular lesson I was anxious to teach him. This, however, meant going through the entire course again! For I wanted Rex to obey my orders even if I did not utter a word—to respond to signals. I evolved a series of signs: a raised finger meant "Sit"; if I ran my finger along the ground it indicated "Track"; my outstretched right hand said "Hand"; and so on. I taught him to understand these signals by first using them together with the spoken word. Then I gave merely the signal, and added the verbal command only if he did not respond. He became so expert that I was later to give fully comprehensive demonstrations with Rex throughout the country—and not utter a single word during the entire performance! I was the first dog handler in the police to introduce this sign language.

Towards the end of his training I took Rex into Esher every day for three weeks, where he put the mid-week shoppers off their stride when we had practice chases among them. Rex bounded round prams, skidded between legs, and slithered round corners, ignoring everything and everyone in his pursuit of the man he was after.

Our great day was July 14, 1950. Training over, our dogs were put through their paces under the eye of Sergeant Boath. Unfortunately, MacCallum was not with us; during the course he had lost Roy because of distemper, and his new charge, Jenny, was not yet fully trained. Although I was supremely confident that Rex would pass without being faulted, I was, at the same time, a little worried and nervous. Suppose (even if the odds were a million to one against), just suppose (despite there being no such possibility) Rex failed? It would mean that Rex and I would be parted, and the prospect was a gloomy one. I am aware that one must have a sense of proportion about animals, that there is something faintly ridiculous, slightly nauseating, in an all-consuming passion for a pet. But dogs have their personalities, idiosyncrasies, and individuality, just as human beings do, and Rex, being so much more intelligent and talented than most, had won my heart. I dreaded the thought of losing him.

Fortunately, there was no need for apprehension. Rex seemed to realize that he was on trial in direct competition with his pals. He sought, tracked, chased, and retrieved with effortless, almost contemptous ease. His responses during the obedience tests were robot-like. All the dogs passed—but it was agreed, unanimously, that Rex was the best of them all. "I told you so," said Tom Scott. "I told you you had a cracker. And you have."

That evening the family were gathered at the front door to hear the news. "Of course he's passed," I said, with an indifference I was far from feeling. "What did you think?"

"I knew he would," said my wife. "I was so sure about it I bought him a special treat."

Rex, sniffing at the kitchen door, already knew. He had a sensitive nose for liver.